Human Reproductive System - Anato

Embryo Development

Stages, Mechanisms and Clinical Outcomes

HUMAN REPRODUCTIVE SYSTEM - ANATOMY, ROLES AND DISORDERS

Additional books in this series can be found on Nova's website under the Series tab.

Additional e-books in this series can be found on Nova's website under the e-book tab.

Human Reproductive System - Anatomy, Roles and Disorders

Embryo Development

Stages, Mechanisms and Clinical Outcomes

**David Reyes
and
Angelica Casales
Editors**

Nova Biomedical

New York

Copyright © 2013 by Nova Science Publishers, Inc.

All rights reserved. No part of this book may be reproduced, stored in a retrieval system or transmitted in any form or by any means: electronic, electrostatic, magnetic, tape, mechanical photocopying, recording or otherwise without the written permission of the Publisher.

For permission to use material from this book please contact us:
Telephone 631-231-7269; Fax 631-231-8175
Web Site: http://www.novapublishers.com

NOTICE TO THE READER

The Publisher has taken reasonable care in the preparation of this book, but makes no expressed or implied warranty of any kind and assumes no responsibility for any errors or omissions. No liability is assumed for incidental or consequential damages in connection with or arising out of information contained in this book. The Publisher shall not be liable for any special, consequential, or exemplary damages resulting, in whole or in part, from the readers' use of, or reliance upon, this material. Any parts of this book based on government reports are so indicated and copyright is claimed for those parts to the extent applicable to compilations of such works.

Independent verification should be sought for any data, advice or recommendations contained in this book. In addition, no responsibility is assumed by the publisher for any injury and/or damage to persons or property arising from any methods, products, instructions, ideas or otherwise contained in this publication.

This publication is designed to provide accurate and authoritative information with regard to the subject matter covered herein. It is sold with the clear understanding that the Publisher is not engaged in rendering legal or any other professional services. If legal or any other expert assistance is required, the services of a competent person should be sought. FROM A DECLARATION OF PARTICIPANTS JOINTLY ADOPTED BY A COMMITTEE OF THE AMERICAN BAR ASSOCIATION AND A COMMITTEE OF PUBLISHERS.

Additional color graphics may be available in the e-book version of this book.

Library of Congress Cataloging-in-Publication Data

ISBN: 978-1-62417-723-1

Library of Congress Control Number: 2012955267

Published by Nova Science Publishers, Inc. † New York

Contents

Preface vii

Chapter 1 Chick Embryogenesis: A Unique Platform to Study the Effects of Environmental Factors on Embryo Development 1
S. Yahav and J. Brake

Chapter 2 Molecular and Cellular Aspects of Blastocyst Dormancy and Reactivation for Implantation 41
Zheng Fu, Yongjie Chen, Weiwei Wu, Shumin Wang, Weixiang Wang, Bingyan Wang and Haibin Wang

Chapter 3 Recent Advances in the Study of Limb Development: The Emergence and Function of the Apical Ectodermal Ridge 77
Joaquin Rodriguez-Leon, Ana Raquel Tomas, Austin Johnson and Yasuhiko Kawakami

Chapter 4 The Use of Time Lapse Photography in an *In Vitro* Fertilization Programme for Better Selection for Embryo Transfer 113
Borut Kovačič, Nina Hojnik and Veljko Vlaisavljević

Index 141

PREFACE

In this book, the authors discuss the stages, mechanisms and clinical outcomes of embryo development. Topics include chick embryogenesis as a unique platform to study the effects of environmental factors on embryo development; molecular and cellular aspects of blastocyst dormancy and reactivation for implantation; limb development and the emergence and function of the apical ectodermal ridge; and the use of time lapse photography in an in vitro fertilization program for better selection of embryo transfer.

Chapter 1 – Bird embryogenesis takes place in a relatively protected environment that can be manipulated especially well in domestic fowl (chickens) where incubation has long been a commercial process. The embryonic developmental process has been shown to begin in the oviduct such that the embryo has attained either the blastodermal and/or gastrulation stage of development at oviposition. Bird embryos can be affected by "maternal effects," and by environmental conditions during the pre-incubation and incubation periods. "Maternal effects" has been described as an evolutionary mechanism that has provided the mother, by hormonal deposition into the yolk, with the potential to proactively influence the development of her progeny by exposing them to her particular hormonal pattern in such a manner as to influence their ability to cope with the expected wide range of environmental conditions that may occur post-hatching. Another important aspect of "maternal effects" is the effect of the maternal nutrient intake on progeny traits. From a commercial broiler chicken production perspective, it has been established that greater cumulative nutrient intake by the hen during her pullet rearing phase prior to photostimulation resulted in faster growing broiler progeny. Generally, maternal effects on progeny, which have both a genetic and an environmental component represented by yolk hormones deposition and embryo nutrient utilization, have an important effect on the development of a wide range of progeny traits. Furthermore, commercial embryo development during pre-incubation storage and incubation, as well as during incubation *per se* has been shown to largely depend upon temperature, while other environmental factors that include egg position during storage, and the amount of H_2O and CO_2 lost by the egg and the subsequent effect on albumen pH and height during storage have become important environmental factors to be considered for successful embryogenesis under commercial conditions. Manipulating environmental temperature during the period of egg storage, during the intermediate pre-incubation period, and incubation period *per se* has been found to significantly affect embryo development, hatching progress, chick quality at hatching, and chick development post-hatching. These temperature manipulations have also been shown to affect the acquisition of thermotolerance to subsequent post-hatching thermal challenge.

This chapter will focus on: a. "maternal effects" on embryo and post-hatching development; b. environmental effects during the post-ovipositional period of egg storage, the intermediate pre-incubation period, and incubation period *per se* on chick embryogenesis and subsequent post-hatching growth and development; and c. effects of temperature manipulations during the pre-incubation and incubation periods on acquisition of thermotolerance and development of secondary sexual characteristics in broiler chickens.

Chapter 2 – Blastocyst activation, a process for the blastocyst to achieve implantation competency is equally important as attainment of uterine receptivity for the success of embryo implantation. While a wide range of regulatory molecules have been identified as essential players in conferring uterine receptivity in both laboratory animal models and humans, it remains largely unknown how blastocysts achieve implantation competency. This chapter will highlight our current knowledge about the mechanisms governing the process of blastocyst activation. A better understanding of this periimplantation event is hoped to alleviate female infertility and help to develop novel contraceptives and new strategies for accessing embryo quality in clinical practice.

Chapter 3 – Vertebrate extremities develop from limb buds, which emerge as paired protrusions in the lateral plate mesoderm. Forelimb buds are located anteriorly and hindlimb buds are positioned posteriorly. The morphogenesis of the limb requires coordinated actions of several organizing centers, among which the apical ectodermal ridge (AER) plays crucial roles in limb development. Recent studies have shown how the life of the AER (induction, maturation, maintenance and regression) is regulated. This regulation includes cell type- and process- specific roles of previously identified molecules, such as fibroblast growth factors (FGFs), Wnts and bone morphogenetic proteins (BMPs). The studies have also revealed several new players, such as Arid3b, R-Spondin 2 and Flrt3. These advances have enhanced the understanding of how the AER is regulated from its emergence to its regression. Progress has also been made in understanding AER function in relation to processes critical for limb development: proximal-distal patterning, anterior-posterior patterning, chondrogenesis and apoptosis. By focusing on two major model systems, chick and mouse embryos, the authors will review recent advances in combination with relevant previous studies in the development and function of the AER.

Chapter 4 – The time lapse photography is not a new method for assessing the dynamics of early embryo development *in vitro*. It has been used many times in the past for studying cleavages and blastulation of embryos of various animal species. However, this technique became available for routine use in an human *in vitro* fertilization (IVF) programme only a couple years ago and it becomes more and more popular today. The new time lapse systems are using modified microscopes which are positioned within the incubators. The observation of embryos does not need the opening of incubators. By sequential photographing of each embryo separately with camera of low intensity illumination, more than 1400 pictures of embryo are made. All these pictures are collected together and transformed into a short movie with software. This system offers the observation of dynamics of embryo development. The studies, which have used a time lapse technique for studying embryo development, revealed that the timing between different events can be used for predicting its developmental potential. In this paper the advantages and drawbacks of time lapse photography is precisely described. An overview through the published papers analyzing the dynamics of human embryo development from the zygote toward blastocyst is done and new timing parameters for grading zygotes, early embryos and blastocysts are analyzed.

In: Embryo Development
Editors: D. Reyes and A. Casales

ISBN: 978-1-62417-723-1
© 2013 Nova Science Publishers, Inc.

Chapter 1

CHICK EMBRYOGENESIS: A UNIQUE PLATFORM TO STUDY THE EFFECTS OF ENVIRONMENTAL FACTORS ON EMBRYO DEVELOPMENT

S. Yahav[*,1] and J. Brake[2]

[1]Institute of Animal Science, ARO the Volcani Center, Bet Dagan, Israel
[2]Department of Poultry Science, NC State University, Raleigh, NC, US

ABSTRACT

Bird embryogenesis takes place in a relatively protected environment that can be manipulated especially well in domestic fowl (chickens) where incubation has long been a commercial process. The embryonic developmental process has been shown to begin in the oviduct such that the embryo has attained either the blastodermal and/or gastrulation stage of development at oviposition. Bird embryos can be affected by "maternal effects," and by environmental conditions during the pre-incubation and incubation periods. "Maternal effects" has been described as an evolutionary mechanism that has provided the mother, by hormonal deposition into the yolk, with the potential to proactively influence the

---

[*] Email: yahavs@agri.huji.ac.il.

development of her progeny by exposing them to her particular hormonal pattern in such a manner as to influence their ability to cope with the expected wide range of environmental conditions that may occur post-hatching. Another important aspect of "maternal effects" is the effect of the maternal nutrient intake on progeny traits. From a commercial broiler chicken production perspective, it has been established that greater cumulative nutrient intake by the hen during her pullet rearing phase prior to photostimulation resulted in faster growing broiler progeny. Generally, maternal effects on progeny, which have both a genetic and an environmental component represented by yolk hormones deposition and embryo nutrient utilization, have an important effect on the development of a wide range of progeny traits. Furthermore, commercial embryo development during pre-incubation storage and incubation, as well as during incubation *per se* has been shown to largely depend upon temperature, while other environmental factors that include egg position during storage, and the amount of H_2O and CO_2 lost by the egg and the subsequent effect on albumen pH and height during storage have become important environmental factors to be considered for successful embryogenesis under commercial conditions. Manipulating environmental temperature during the period of egg storage, during the intermediate pre-incubation period, and incubation period *per se* has been found to significantly affect embryo development, hatching progress, chick quality at hatching, and chick development post-hatching. These temperature manipulations have also been shown to affect the acquisition of thermotolerance to subsequent post-hatching thermal challenge.

This chapter will focus on: a. "maternal effects" on embryo and post-hatching development; b. environmental effects during the post-ovipositional period of egg storage, the intermediate pre-incubation period, and incubation period *per se* on chick embryogenesis and subsequent post-hatching growth and development; and c. effects of temperature manipulations during the pre-incubation and incubation periods on acquisition of thermotolerance and development of secondary sexual characteristics in broiler chickens.

INTRODUCTION

Developmental biology has long taken advantage of oviparous species to access embryos at early developmental stages and study embryo maturation ex utero. Among the models, the domestic chick embryo has provided important insights into development and organogenesis. Chick embryos have been well characterized and often utilized to elucidate developmental mechanisms during embryogenesis, as well as to explore candidate gene expression. These embryos have also become an established model for tissue/cell transplantation because of the lack of a mature immune system during early development. The chick embryo has also become recognized as a model for mammalian stem cell and cancer research because of ease of access for various manipulations and injections of different substrates.

The avian egg represents a fundamental reproductive strategy in colonization of terrestrial habitat. The amniotic egg has the novel presence of three extra-embryonic membranes, amnion, chorion, and allantois, which surround the embryo, mediate between it and the environment, and provide an aqueous medium in which to complete development and ontogenetic programming inside the eggshell.

In the avian class, the chick embryo provides a wide range of accessible and economical in vivo models that may be used to study "maternal effects", fertile egg storage conditions, and potential manipulations during embryogenesis that affect not only embryo development, but also post-hatching parameters that are especially important for domestic fowl undergoing genetic selection for specific economically important traits. The unique situation of domestic fowl can be best illustrated by broiler chickens that have been subject to an exceptional situation of intensive genetic selection for many years. Recent decades have seen significant progress in the genetic selection of the meat-type fowl, i.e., broilers (Havenstein et al., 1994ab, 2003ab) that has led to rapid growth accompanied by increased feed efficiency and metabolic rate (Janke et al., 2004, Druyan 2010), with consequently elevated internal heat production (Sandercock et al., 1995). Such developments should logically require parallel increases in the size of the cardiovascular and respiratory systems, as well as enhancements in their functional efficiency. However, inferior development of such major systems (Havenstein et al., 2003b) has led to relatively low capability to maintain adequate dynamic steady-state mechanisms in the body, which should balance energy expenditure and body water balance under extreme environmental conditions (Yahav, 2009).

The consequences of broiler genetic selection starts in the embryo (Hulet, 2007), which represents 30-40% of the total broiler life cycle inside the egg during incubation, as well as up to 14 days in storage prior to incubation (Fasenko, 2007). Thus, anything that can optimize growth and development post-hatching by altering what has often been thought up until the present time as the need for relatively constant environmental conditions during pre-incubation and incubation may be of great interest. Moreover, the knowledge of how to influence the maternal contribution to progeny development through fertile egg management and incubation may also be of great interest.

Embryo development during the storage period between oviposition and incubation largely depends upon temperature (Brake et al., 1997; Fasenko, 2007). Other environmental factors that include egg position (large or small end up; Elibol et al., 2002), and the amount of H_2O and CO_2 lost by the egg as they affect albumen pH and height during storage (Meijerhof, 1992; Brake et al., 1997; Güçbilmez et al., 2010) have also an influence on the development of the embryo. However, there has been no doubt that temperature continues to play a major role in embryo development during incubation (Yahav et al., 2009; Piestun et al., 2009), while relative humidity (Meijerhof, 1992), CO_2, and hypoxia (Druyan et al., 2012) have measureable but less pronounced effect on the developmental process.

During the past few decades the importance of maternal effects on embryo development and post-hatching progeny performance has received increased attention as well. Maternal effects have been defined as phenotypic variation in progeny that was a consequence of the maternal phenotype (Roff, 1998). Maternal effects have both genetic and environmental components, enabling adaptive responses of the progeny in heterogeneous environments (Mousseau and Fox, 1998). Maternal effects on progeny that have both genetic and environmental components have been shown to be elicited by yolk hormone transfer and embryo nutrient utilization that have important effects on the development of a wide range of progeny traits (Gil, 2008; Groothuis and Swabl, 2008; Isakson et al., 2010; Müller et al., 2012) and may be significantly affected by the aforementioned egg storage and incubation parameters.

Apparent sex ratio can also be affected by maternal effects. Birds exhibit genetic sex determination (Smith and Sinclair, 2004) where progeny sex is decided at fertilization by heritable sex chromosomes that are ZZ for males and ZW for females. However, the apparent sex ratio at hatching has been found to be affected by yolk hormones (Love et al., 2005; Hayward et al., 2004; Rubolini et al., 2006; Rutkowska and Badyaev, 2008), either by affecting meiosis and thereby sex determination or by affecting resorption or growth of the follicle of the opposite sex (Alonso-Alvarez, 2006), which has provided the possibility of maternal modification of secondary sexual characteristics or even primary sex ratio. Importantly, several studies have demonstrated that elevation of maternal testosterone induced a male biased sex ratio (Veiga et al., 2004; Rutkowska and Cichon, 2006; Isaksson et al., 2010). This evolutionary development has provided the hen with the potential to proactively influence the development of her progeny by exposing them to her particular hormonal pattern in such a manner as to presumably influence their ability to cope with a wide range of environmental conditions post-hatching.

Another important aspect has been the effect of maternal nutrient intake on progeny traits. It has been found that greater cumulative nutrient intake by the hen during her pullet rearing phase prior to photostimulation resulted in faster growing broiler progeny (Brake et al., 2004), which suggested that hens that could produce chicks with greater broiler potential would express that progeny potential if provided sufficient nutrition during rearing. While this was certainly not the exact scenario that would be experienced in feral conditions, it could be suggested that the innate system of hormonally-influenced transfer of information via the yolk evident in feral birds may have been so modified through modern genetic selection and management systems. However, the mechanism by which, or to what extent the hen controls the transfer of hormones into her eggs and the metabolic consequences of this (Groothuis and Swabl, 2008) has not been fully characterized. However, it has been reported that yolk sac absorption and utilization was decreased by increasing egg storage time that probably influenced both nutrient utilization and the potential influence of yolk hormones, as evidenced by decreasing broiler performance with increasing length of fertile egg storage (Afsar et al., 2007). Although important components of nutritional and hormonal programming of the developing embryo and post-hatching chick may lie within the yolk, the thickness and pH of the albumen, and egg surface area (respiratory pore area) relative to egg volume (embryo weight) that collectively determine vital gas exchange potential (conductance) of the egg obviously interact to help determine yolk absorption and utilization. Nevertheless, it has become clear that the interaction among maternal effects and incubation factors play a crucial role in the initiation of embryo development, mortality of embryos, length of incubation, absorption of yolk, hatchability, and post-hatching broiler performance.

This chapter will focus on: a. "maternal effects" on embryo and post-hatching development; b. environmental effects during the post-ovipositional period of egg storage, the intermediate pre-incubation period, and incubation period *per se* on chick embryogenesis and subsequent post-hatching growth and development; and c. effects of temperature manipulations during the pre-incubation and incubation periods on acquisition of thermotolerance and development of secondary sexual characteristics in broiler chickens.

MATERNAL EFFECTS

"Maternal effects" has been defined as phenotypic variation in progeny that was a consequence of the phenotype of the mother (Roff, 1998). These effects occur when progeny phenotype was influenced by environmental variables experienced by the hen (Mousseau and Fox, 1998), enabling an adaptive response by the progeny to a heterogeneous environment, as a result of the potential of these effects to generate immediate phenotypic changes via phenotypic plasticity (Räsanen and Kruuk, 2007). It has been well documented that hens can influence their progeny in a non-genetic manner by allocating resources differentially in a context-dependent way, whereas factors such as diet, climatic conditions, and mate quality may affect progeny through a maternal influence (Gilbert et al., 2007).

Female birds present an extraordinary opportunity to study the implications of maternal effects because their eggs contain a complex of resources for the embryo that can be easily measured and experimentally modified. A variety of egg components such as egg mass, hormones, antioxidants, and immune factors has been shown to influence different progeny traits (Schwabl, 1993; Gasparini et al., 2001; Saino et al., 2003; Biard et al., 2005). The most interesting parameter has been the hormonal one, playing a major role in organizing phenotypic differentiation and regulating physiological functions (Nelson, 2000). Among endocrine effects, steroid hormones have received the greatest attention recently. These hormones have been shown to be transferred in substantial amounts into the yolk and early exposure of the embryo to these hormones has been shown to strongly affect progeny survival, behavior, morphology, physiology, and even sex (Groothuis et al., 2005). Although exposure of an embryo to maternal steroids has been viewed as a potentially detrimental perturbation of development (Carere and Balthazart, 2007), research during the last decade has suggested that such hormone mediated maternal effects may be adaptive (Groothuis et al., 2005).

Environmental and Nutritional Modification of Maternal Effects

Individuals of the same animal species and sex may differ in morphology and physiology, as well as behavior. In many cases, the developmental outcome may be influenced by a prediction of the conditions in which the new individual will subsequently live. Parents provide a conduit by which environmental signals can be passed to progeny (Groothuis et al., 2005).

Maternal effects can be exerted through a diversity of pathways and at different periods during the life cycle, however, the prenatal period may be the most influential for two reasons: a. It has been well established that early environmental influences lead to irreversible modifications, and the embryo must be reliant upon environmental information provided by the mother as the embryo clearly has no contact with the environment, especially in domestic fowl where incubation has typically been carried out under relatively constant machine conditions; b. Gonadal hormones have been described as the probable maternal to embryo mediators. Early exposure to certain hormones resulted in long-lasting organizing effects on brain and behavior (Ryan and Vandenbergh, 2002), which may lead to expression of different phenotypes within the same sex. The avian female has been shown to be able to affect the development of its progeny both genetically and by transfer of hormones that reflect the impact of environmental stimuli through an epigenetic mechanism (Groothuis et al., 2005; Mousseau and Fox, 1998; Müller et al., 2012). Although the environment of commercial poultry has become relatively constant due to modern housing technology, it has been reported that greater cumulative nutrient intake by the hen during her pullet rearing phase prior to photostimulation and sexual maturity resulted in faster growing broiler progeny (Brake et al., 2004) as shown in Figure 1 and Table 1. All males that were mated to these hens were grown sex-separate as a single group to a cumulative nutrient intake of about 32,000 kcal metabolizable energy and 1600 grams crude protein known to be sufficient for optimum reproductive function.

This was a somewhat dose-dependent effect with greater cumulative pullet nutrition associated with greater broiler growth in a linear manner, which suggested that hens that could produce chicks with greater broiler potential would do so if provided sufficient nutrition during rearing. This was subsequently confirmed by Romero-Sanchez et al. (2007) with broilers grown to 49 days of age. While certainly not the exact scenario that would be experienced in feral conditions, these data did suggest that eggs produced from relatively more nutrient-replete hens produced broiler chicks prepared to grow in a nutrient-replete environment. It was suggested that the innate system of hormonally-influenced transfer of information and nutrients via the yolk that was evident in feral birds may have been modified into this scheme through modern genetic selection and management. This scheme was found to be modified by the number of hens per cage (1 versus 2) that created differential competition for feed and space as well as altered levels of stress (Eusebio-Balcazar, 2009) during the critical weeks following photostimulation.

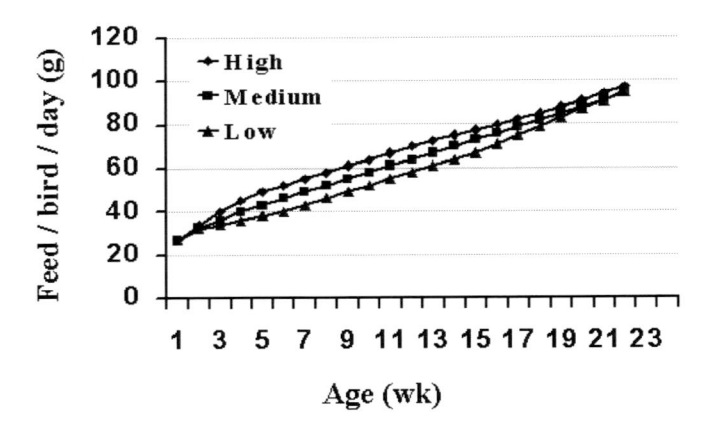

Figure 1. Pullet feeding programs used to produce three graded levels of cumulative crude protein (CP) and metabolizable energy (ME) intakes (High (27,788 kcal ME and 1,485 g CP), Medium (26,020 kcal ME and 1,391 g CP), Low (24,242 kcal ME and 1,296 g) for broiler breeder (broiler parent) hens prior to sexual maturity. (After Brake et al., 2004).

Table 1. Effect of cumulative rearing nutrition prior to sexual maturity of broiler breeder hens (female broiler parent) on subsequent broiler progeny body weight from eggs laid at 28 weeks of age (after Brake et al., 2004)

Breeder Parent Age (weeks)	Broiler Sex	Cumulative Rearing Nutrition[1]			P =
		High	Medium	Low	
		------------------(g)------------------			
28	Male	946	935	901	0.06
28	Female	896	873	859	0.03

[1] Pullet feeding programs as shown in Figure 1 that were used to produce three graded levels of cumulative crude protein (CP) and metabolizable energy (ME) intakes (High (27,788 kcal ME and 1,485 g CP), Medium (26,020 kcal ME and 1,391 g CP), Low (24,242 kcal ME and 1,296 g) for broiler breeder (broiler parent) hens prior to sexual maturity. Males that had been reared separately on a single adequate rearing program were mated to these females at 22 weeks of age, when photostimulation occurred. The start of significant egg production was at 25 weeks of age.

This latter work demonstrated that cumulative feeding program during rearing and altered competition at photostimulation changed the vital gas exchange potential (conductance) of the egg and affected progeny leg development in such a manner as to affect broiler walking ability, which would certainly impact feed intake and therefore growth and feed efficiency. In a cage study, a reduced rate of feed increase from 5% rate of lay to peak rate of lay did not affect egg weight or percentage shell but did reduce albumen height that probably promoted improved vital gas exchange during early incubation (Peebles et al., 2000) as evidenced by larger chicks from similar weight eggs followed by a 156 g greater broiler body weight at 42 days of age (Leksrisompong, 2010). Thus, the evolutionary scheme has provided the hen with the potential to adaptively adjust the development of her progeny by exposing them to her nutritional scheme in a manner that improved their growth performance. Such exposure would certainly be influenced in the domestic fowl by management factors that influence yolk sac absorption during incubation, an issue central to "chick quality" that has been frequently discussed within the commercial poultry industry.

Sex Determination

Sex determination has been described as a process of great antiquity evident in animals from simple eukaryotes to mammals involving genetic and/or environmental factors. Birds exhibit genetic sex determination (Smith and Sinclair, 2004) where sex has been decided at fertilization by the inheritable sex chromosomes that have been termed ZZ for males and ZW for females. However, it has not been determined if Z chromosome dosage, or if the W chromosome carried a dominant ovary determinant, or a combination of the two, were most important. The male candidate gene *DMRT1* involved in sex determination has been located on chromosome Z. Possible ovary determinant genes located on chromosome W include *FET1* and *ASW* (Smith and Sinclair, 2004; Zhao et al., 2010). Although it has remained unclear how sex chromosomes ultimately determine the phenotypic sex (Smith et al., 2007), the cellular basis of sexual differentiation has been demonstrated. An early event in sexual differentiation of chickens was reported to be the differential expression of the enzyme aromatase in the genital ridge, where female embryos express more aromatase than males as early as E6.5 (Yoshida et al., 1996; Shimada, 1998) but the molecular signal that controlled the early expression of aromatase in females has not been identified (Nishikimi et al.,

2000; Balthazart et al., 2009). Aromatase has been reported to catalyze the transformation of androgens such as testosterone (T) or androstendione (A4) into estrogens such as 17β estradiol (E2) or estrone (Lephart, 1996).

Maternal and Environmental Effects on Sex Determination

Sex determining genes become active during early embryogenesis. Gonadal development consists of primordial gonads that develop from the intermediate mesoderm on the ventromedial surface of the embryonic kidney. By E3.5 the gonadal cortex has developed, after which the medullary cords appear beneath the cortex. The embryo gonads remain indifferent from E3.5 to E4.5, being morphologically similar in both genders. The primordial germ cells (PGCs) then migrate via the bloodstream to the gonadal cortex or medulla depending upon which gonadal portion has proliferated to claim the PGCs. Proliferation and claiming of PGCs by the cortex has been demonstrated to produce an ovary while claiming by the medulla produced a testis. Consequently, gonadal differentiation has become clear by E6.5 (Ebensperger et al., 1988). While in mammals gonadal hormone synthesis and secretion was shown to occur after sex determination, in birds gonadal development was susceptible to hormonal manipulation, and phenotypic sex reversal could be induced by estrogen injection. In such cases where the avian embryo was susceptible to hormonal manipulation, the question of the functional significance of sex steroid hormones that originated from a maternal source and were presented via the egg yolk (and affected by rate of yolk sac absorption) would be of great interest. Gonadal development, which was susceptible to hormonal manipulation in birds, has provided an opportunity for estrogen to affect sex determination (Scheib, 1983). High concentrations of androgens and estrogens (testosterone (T), 5-dihydrotestosterone (DHT), androstendione (A4), and 17β estradiol (E2)) were found in the egg yolk of chickens (Eising et al., 2003; Elf and Fivizzani, 2002) as well as in yolks of other avian species in much higher concentrations than in the embryo or adult (Groothuis and Swabl, 2002; Groothuis et al., 2005). These steroids were evidently deposited by the laying female, being detected in the yolk of both freshly laid infertile and fertile eggs, before an embryo could have produced these steroids. The concentration of these hormones varied in relation to environmental variables such as intra and/or intersexual competition, food availability, mate activity, position in clutch, level of stress, and environmental conditions (Groothuis and Schwabl, 2002; Groothuis et al., 2005; Schwabl,

1993). This could be surmised to be similar to the nutritional and management scenario described above for broiler breeder pullets and hens. Injection of T and A4 into eggs in order to mimic a natural endocrine manipulation, within the highest physiological levels, produced effects that included earlier hatching time, increased development of the piping muscle, and increased growth (Eising et al., 2006; Sockman and Schwabl, 2000). Moreover, Müller et al. (2012) suggested that T concentration in the freshly laid eggs of Fife Fancy Canaries was changed via changes in yolk mass and position in the clutch. It has been suggested that these hormonal insertions may have long-lasting effects that permanently shaped the phenotype (Eising et al., 2006; Rubolini et al., 2006).

In many avian species males and females respond differentially to gonadal steroids, which largely explain sex-specific behavioral phenotypes. This sex difference explains the effect of embryonic exposure to steroid hormones. Steroids exert their actions early in life during a *short critical period* required to organize in an irreversible manner the responsiveness to sex steroids and the underlying neural substrate (Carere and Balthazart, 2007). In the domestic chick, injection of estrogen into an egg that contained a male embryo demasculinized the male such that male-typical copulatory behavior in response to T post-hatching was not evident (Balthazart and Ball, 1995). On the other hand, exposure to ovarian estrogens demasculinized females during embryogenesis. Although sex steroids have been shown to activate the expression of sexual behavior in adult birds, this sexual expression was suppressed by the effect of steroid hormones that permeate the neural substrate during ontogeny. Sexual differentiation has been found to be influenced by temperature during embryonic development in many species including fishes (Oldfield, 2005; Piferrer et al., 2005), amphibians (Eggert, 2004; Sakata et al., 2005), and reptiles (Bull, 1980; Ewert and Nelson, 1991; Pieau et al., 1999) in a process known as temperature dependent sex determination. In some amphibians (Eggert, 2004; Sakata et al., 2005) and lizards (Shine et al., 2002; Quinn et al., 2007), extreme temperatures during embryogenesis overrode genetic sex determination in a process known as sex reversal. A study with chickens (Ferguson, 1994) demonstrated that pulsing eggs with abnormally high or low temperatures produced phenotypic sex reversal in 10% of the hatchlings. Recent studies of Göth and Booth (2005) and Göth (2007) with the Australian brush-turkey showed that the first two weeks of embryogenesis was a thermally sensitive period, during which high and low incubation temperatures resulted in more genotypic females and males, respectively. Eiby et al. (2008) subsequently showed the mechanism for altered sex ratio to be temperature-dependent sex-biased embryo mortality and not sex reversal as in fishes, amphibians, and reptiles. However, sex-biased embryo mortality would not be of interest in commercial broiler scenarios where high chick hatchability was desired.

Hormone-Mediated Maternal Effects

Maternal strategies and the effect that they have on their progeny can be adaptive irrespective of whether they apparently help or hinder their own individual progeny (Russell and Lummaa, 2009). The environment provided by the mother must be among the most significant that progeny experience during development. Maternal effects in birds have been reported to provide epigenetic modifications of progeny phenotype relative to the environment provided by the hen during development (Mousseau and Fox, 1998). Maternal effects provide the possibility that hens may adjust the development and phenotype of their progeny according to environmental conditions, using steroid hormones that exploit the adaptive phenotypic plasticity approach. Steroids have been shown to be integral regulators evoking a cascade of programmed processes of development and differentiation leading from the genotype to phenotypic manifestation. Furthermore, steroids have also been found to affect phenotypic responses to environmental changes throughout life. Schwabl (1993) was the first to show that T was found in Canary eggs. The T concentration in fresh eggs of the canary increased within the clutch with order of lay, irrespective of the genetic sex of the progeny. The finding that sex differences may be affected by yolk hormones (Love et al., 2005; Hayward et al., 2004; Rubolini et al., 2006; Rutkowska and Badyaev, 2008) has opened the possibility of maternal modifications of secondary sexual characteristics under commercial conditions.

ENVIRONMENTAL EFFECTS

In nature, incubation conditions have been reported to be clearly non-uniform, because of the need of the parents to search for food and escape from predators, and because of non-uniform nest insulation (Webb, 1987). This has contradicted the relative environmental uniformity under which commercial broiler eggs have been typically stored and incubated. This may be one of the reasons why birds in the wild have become quite capable of coping with extreme environmental conditions, a trait that has somewhat deteriorated along with advanced genetic selection in broilers.

Environmental Effects during Fertile Egg Storage

Embryonic developmental processes have been demonstrated to begin with fertilization and continue during the completion of egg formation in the oviduct of the hen. An embryo has been reported to comprise 40,000 to 60,000 cells (Eyal-Giladi and Kochav, 1976) at oviposition. Albumen quality has also been reported to begin to diminish prior to oviposition presumably due to embryonic production of ammonia (Benton and Brake, 2001).

The progression of embryo development during pre-incubation storage has been related to environmental factors such as temperature, relative humidity (RH), ammonia, CO_2, and others. The relatively reduced rate of embryo development during storage largely depends upon temperature (Brake et al., 1997; Fasenko, 2007), which can range between 14 and 21°C (Fasenko et al., 1992), while optimal hatchability after long term storage (>14 days) was achieved when the storage temperature was 12°C (Funk and Forward, 1960), but 15°C was better for eggs stored for 8 days, and 18°C was best for eggs stored for 2 days (Kirk et al., 1980). These data simply reflected the effects of storage period appropriate temperature on albumen quality. The discrepancy among studies may be also caused by different tissues of the embryo having different temperature requirements and the need to reduce embryo metabolism relative to length of storage. Fasenko et al. (2002) demonstrated differences in embryonic metabolic rate post-storage between eggs that had been stored for 4 and 15 days. This study and that of Christensen et al. (2001) provided evidence that embryos from long-stored eggs not only lagged in development, but their metabolic rate during subsequent incubation was also changed by storage. Storage temperature may also help control embryonic production of ammonia that has to be absorbed by the albumen (Benton and Brake, 2001). Storage of eggs in the small end up position may diminish potentially harmful effects of ammonia (Elibol et al., 2002), probably by positioning more albumen near the blastoderm during extended periods of storage. Other issues that have been raised concerning egg storage involve the initiation of incubation and response to incubation temperature by specific tissues. A longer storage period was reported to cause a delay in the initiation of the development process (Arora and Kosin, 1966) and the process was at a slower rate during the initial portion of incubation (Mather and Laughlin, 1977;

Meijerhof, 1992). As an example, under identical incubation and egg storage conditions chicks that hatched from eggs stored at constant conditions for 10 days versus 1 day exhibited greater residual yolk sac weight as well as reduced liver and gizzard weight (Afsar et al., 2007). Further, body weight at 42 days of age was decreased due to storage by 58 g or 138 g when the eggs were produced by 34 or 59-week-old broiler breeder flocks, respectively (Ates et al., 2004). These data demonstrated that long egg storage somehow interfered with critical aspects of yolk sac absorption that adversely affected nutrient and/or hormone assimilation into the developing embryo and hatched chick, and influenced broiler progeny growth. This problem may be addressed by a more rapid increase in egg temperature during preheating prior to incubation that was reported to decrease early embryo mortality (Güçbilmez et al., 2009).

Water loss by evaporation through the eggshell during egg storage has been shown to be influenced by RH, temperature, shell porosity, and air movement (Spotila et al., 1981). It has been stated that high RH must be maintained during storage in order to prevent as much water loss as possible from the egg in order to maintain maximal hatchability (Proudfoot, 1976; Mayes and Takeballi, 1984; Meijerhof, 1992). However, Walsh et al. (1995) concluded that only eggs from older flocks with poorer albumen quality were very sensitive to lower RH. Weight loss from the egg during storage has been observed to proceed in a somewhat linear manner for about 7 days after which a plateau was observed. This was interesting as it has been generally accepted that broiler performance will decline when eggs were stored for more than 7 days. Weight of the yolk has also been observed to increase during storage, which indicated that water was moving from the albumen into the yolk as well as through the egg shell. There has been evidence developed that during heating of eggs to 37.5°C for 3 hours during long storage, yolk weight declined and albumen height recovered slightly (French, personal communication, 2011). Heating of eggs during 11 days of storage returned hatchability to near short-term storage levels (Güçbilmez et al., 2010). These data may provide evidence of a reversal of water movement that helped re-establish the necessary yolk-albumen gradient across the blastoderm as discussed below.

The main effect of CO_2 loss in association with storage has been related to albumen pH that increased from 7.6 at oviposition to as much as 9.0 to 9.5 with long term storage (Goodrum et al., 1989; Stern, 1991, Benton and Brake, 2001). The yolk pH remained fairly stable at approximately 6.0. The establishment and maintenance of an appropriate pH gradient between the albumen and the yolk (across the blastoderm) appeared to be necessary for optimum embryogenesis (Stern, 1991). The albumen pH for optimal embryogenesis has been estimated to be between 8.2 and 8.8 (Walsh et al., 1995). A prolonged increase in albumen pH prior to incubation negatively affected hatchability while it was further hypothesized that an optimal albumen quality as well as optimum albumen pH was necessary prior to incubation (Walsh et al., 1995). It was previously hypothesized that holding eggs in the small end up position may positively affect the pH gradient between the albumen and yolk.

Turning fertile eggs during storage was not recommended prior to 1950. Several studies have since found that the success of turning depended upon storage period and flock age, which suggested an albumen quality relationship. Funk and Forward (1960) recommended turning of eggs stored for 11 to 14 days. Subsequent work showed that a 90° turning treatment exhibited higher hatchability compared to no turning (Becker et al., 1969). It was also shown that eggs from old flocks were more sensitive to insufficient turning than eggs from young flocks, presumably due to differences in albumen quality. Further, storage of eggs in the small end up position was equivalent to turning during storage and more easily applied (Elibol et al., 2002). Both of these methods maintain the embryo positioned on the surface of the yolk in contact with relatively more "fresh" albumen that would absorb ammonia more easily.

Preliminary data from a current study also showed that turning eggs 96 times daily versus 24 times daily during incubation after only 2 days of storage produced significantly longer chicks with smaller yolk sacs at hatching (Lin et al., 2010). The suggested mechanism was through enhanced development of the vascularization of the yolk sac membrane (Lin, 2011), which could be postulated to enhance absorption of nutrients and/or hormones deposited therein. Indeed, turning 96 times daily was previously found to be of greater significance with long-stored eggs (Elibol and Brake, 2008).

Collectively, there is no doubt that the interaction between egg storage period, storage temperature, storage position, and turning during incubation all have crucial roles in the initiation of embryo development, mortality of embryos, the length of incubation, yolk sac absorption, hatchability, and quality of the newly hatched chick.

Increased Temperature Prior to Incubation – Pre-Incubation

In contrast to natural conditions, commercial fertile eggs have been generally stored in environmental temperatures that ranged from 14 to 21°C and for 1 to 14 days (Fasenko et al., 2002; Meijerhof, 1992). The optimum temperature for subsequent incubation has been reported to be from 37.0°C to 38.0°C for chicken eggs (Insko, 1949; Romanoff, 1960; Landauer, 1967; Lundy, 1969; Wilson, 1991). There has been some concern that the significant temperature gradient between storage and incubation may cause "temperature shock" to be experienced by the embryo, which may negatively affect development. Another concern has been the non-uniformity of embryogenesis imposed by less than uniform rates of warming.

Development of the embryo has been shown to begin at the time of fertilization while the egg was still being formed in the oviduct of the hen such that an embryo in the process of active hypoblast formation containing up to 60,000 cells was present at oviposition (Rudnick, 1944; Eyal-Giladi and Kochav, 1976). Hays and Nicolaides (1934) concluded that pregastrula and early gastrula were the most common stages of development in freshly laid eggs for young and old flock eggs, respectively. A well-advanced gastrula has been most commonly found in eggs from prime flock (middle of laying cycle) eggs (Taylor and Gunns, 1939). Due to the fact that eggs have been found to be in different developmental stages at the time of oviposition, pre-incubation has become a part of hatchery management as pre-incubation has provided a means to incrementally increase the temperature of eggs just prior to incubation in order to improve uniformity of embryogenesis. Furthermore, accumulated data demonstrated that extended egg storage somehow interfered with critical aspects of yolk sac absorption that adversely affected nutrient and/or hormone assimilation into the developing embryo and hatched chick, which negatively influenced broiler progeny growth. However, researchers have warned that pre-incubation can be harmful to some freshly laid eggs due to advancing eggs that were laid in an "optimal" stage to a "less-optimal" stage during and after the pre-incubation treatment (Kosin and Pierre, 1956). On the other hand, Elibol and Brake (2008) demonstrated that employing a more rapid increase in egg temperature prior to incubation reduced early embryo mortality in larger eggs or in younger flock eggs (Güçbilmez et al., 2009). Eggs being pre-incubated that experienced a high air velocity were warmed rapidly, while eggs at a low air velocity required several hours to warm (Reijrink, 2010). Therefore, pre-incubation of eggs prior to incubation has been generally found to be beneficial as demonstrated by an increased hatchability of chicken and turkey eggs (Kosin and Pierre, 1956; Becker and Bearse, 1958). Another goal of pre-incubation prior to incubation has been to reduce the "temperature shock" experienced by the embryo moving from low temperature storage to the higher temperatures of an incubator by allowing the eggs to warm up to an intermediate temperature before setting in the incubators (Renema et al., 2006; Brannan, 2008). By increasing the egg temperature to an intermediate level, the eggs were thereafter able to achieve their incubation temperature more rapidly when set in an incubator. This has been suggested to promote early embryonic growth. Embryos from heavily selected broiler strains, such as those found in the majority of commercial broiler hatcheries worldwide, have been demonstrated to be intolerant of temperature variations with abnormalities and mortality of the embryo being the penalties for exceeding the narrow temperature range that has been thought to be optimum for incubation (Wilson, 1991; Decuypere and Michels, 1992).

Increased temperature during pre-incubation (Piestun et al., 2012) of broiler eggs that had been stored for 4 or 9 days produced accelerated hatching in such a manner that 50% hatchability for eggs that had been subject to storage for 4 days was earlier by 8 hours. For eggs that had been subjected to 9 days of storage, hatching was 4 hours earlier due to the pre-incubation treatment. With eggs stored for 4 days, pre-incubation elevated hatchability by 10%, whereas with eggs stored for 9 days the effect was much smaller, 2% only. This study suggested that pre-incubation had a positive and pronounced effect on incubation duration and hatchability when storage time was for 4 days but this effect declined with a storage period of 9 days. The hatched chicks were thereafter raised during winter and summer (Piestun et al., 2012). In both seasons a positive effect on body weight at marketing age (35 days of age) was observed (Table 2).

This was coupled with effects on several secondary sexual phenotypic characteristics including relative weights of the comb, wattles, and testes (Table 2). The changes in the different secondary sexual characteristics of males and females exposed to pre-incubation treatment seemed to be related to the pre-incubational effect on plasma testosterone concentration that was found to be higher in the growing broilers. Increased plasma testosterone probably also explained the increased breast muscle that was elicited by pre-incubation. However, this phenomenon has to be more carefully studied in order to fully elucidate the complex mechanism that must involve much more than only testosterone.

Conclusively, pre-incubation of embryos prior to incubation has been found to reduce the "temperature shock" due to the gradient of temperatures between storage and incubation, improve the uniformity of embryo development, improve hatchability, and improve growth and development post-hatching.

Table 2. The effect of pre-incubation (Pre[1]) as compared to no pre-incubation (Con) on body weight and secondary sexual phenotypic characteristics of broiler males and females raised in winter or summer environmental conditions (according to Piestun et al., 2012)

Variables	Winter Experiment (4 days storage)				Summer Experiment (9 days storage)			
	Males		Females		Males		Females	
	Con	Pre	Con	Pre	Con	Pre	Con	Pre
Body weight at 35 days (g)	2071	2103	1781[b]	1830[a]	2080[b]	2132[a]	1765[b]	1789[a]
Breast muscle (g/100g body weight)	19.1	19.1	18.9[b]	19.6[a]	19.2	19.4	19.2[b]	20.0[a]
Comb (mg/100g body weight)	35[b]	46[a]	13[b]	14	39[b]	47[a]	13[b]	16[a]
Wattles (mg/100g body weight)	37[b]	44[a]	24	25	51	54	26[b]	31[a]
Testes (mg/100g body weight)	15[b]	18[a]	-	-	17[b]	20[a]	-	-
Ovary (mg/100g body weight)	-	-	20	20	-	-	23	22

[a,b] Within each experiment and gender, means that possess different superscripts differ significantly ($P \leq 0.05$).
[1] Control eggs were not subjected to pre-incubation treatment and were moved from an egg cooler at 18°C and directly to an incubator at 37.5°C. The Pre treatment eggs were moved from an egg cooler and subjected to a pre-incubation treatment of 12 h at 30.2°C before being incubated as the Control.

Increased Temperature during Incubation – Thermal Manipulations (TM)

The optimum incubation temperature for chicken eggs has been reported to be between 37.0°C and 38.0°C (Insko, 1949; Romanoff, 1960; Landauer, 1967; Lundy, 1969; Wilson, 1991). This narrow range of temperature has been adopted in commercial incubation equipment in order to achieve optimal embryo development coupled with maximum hatchability and chick quality. However, the long-standing paradigm of constant incubation conditions has changed in recent years as efforts to develop an enhanced thermotolerance capacity on the part of broilers has been pursued (Yahav, 2009).

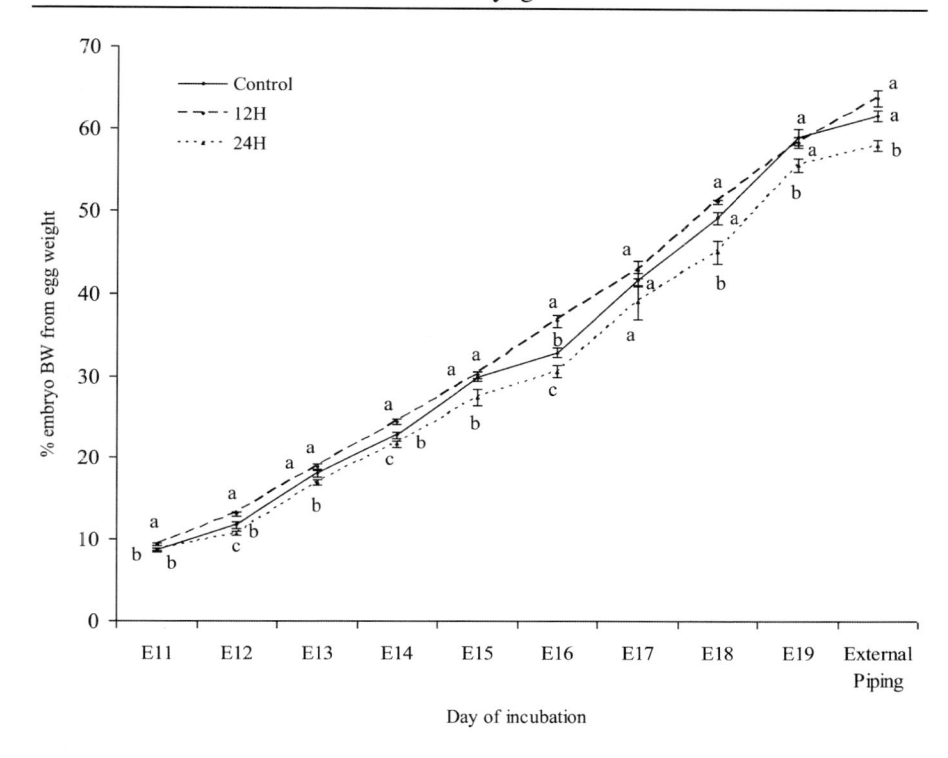

Figure 2. Embryo weight as percentage of egg weight incubated under control conditions (37.8°C and 56% RH) or thermal manipulated (TM) from E7 to E16 inclusive at 39.5°C and 65% RH continuously (24H treatment) or intermittently (12H treatment). Within each day of incubation, different letters indicate significant differences ($P \leq 0.05$) among treatments. n=10. (According to Piestun et al., 2009).

Piestun et al. (2009) developed a thermal manipulation (TM) model for broiler embryos that caused a significant improvement in the acquisition of thermotolerance by post-hatching broilers. The model was based on elevation of incubation temperature to 39.5°C for 12 hours/day or 24 hours/day from E7 to E16 inclusive.

It has been previously reported that elevation of temperature at a very early stage of incubation accelerated growth and development (Romanoff, 1960; Ricklefs, 1987) but had a negative effect on body weight at hatching. Chicken embryos develop during the first 10 days of incubation and thereafter they mainly grow. In the study of Piestun et al. (2009) it was clearly observed that 24 hours/day TM during the second half of embryogenesis negatively affected development leading to a significantly reduced body weight of the hatched chicks (Figure 2). This was consistent with other reports of accelerated development and decreased body weight at hatching as a result of elevated incubation temperature (Webb, 1978; Leksrisompong et al., 2007). Moreover, other studies have demonstrated that TM during mid-incubation also influenced organ weights in the developing embryo and at hatching (Yalcin et al., 2008). One of the organs that was decreased in weight as a result of the 24 hours/day TM was the liver. Not only was the weight of the liver affected, but also its glycogen content was lower. Reduced liver glycogen content has been associated with depressed embryo survival (Christensen et al., 1993; 1999). Indeed, the lower glycogen content of the liver and the piping muscle of the 24 hours/day TM embryos prior to hatching significantly affected their ability to complete hatching (Table 3), which dramatically reduced the overall hatchability of embryos treated in this manner. The most remarkable characteristics of the hatching chicks from the 24 hours/day TM were poor yolk sac absorption and increased evidence of rough and unhealed navels (Table 3) as well as their dull white color and short down (initial feathers). This was in agreement with the findings of Leksrisompong et al. (2007), who reported white-colored chicks with a generally abnormal and unhealthy appearance that might have been due to poor absorption of the yolk sack pigments as a result of an elevated incubation temperature.

In contrast to the 24 hours/day TM, the intermittent TM (12 hours/day) did not negatively affect embryo growth having a positive effect on the embryo relative weight on a number of days during TM (Figure 2). This agreed with the findings of Yalcin et al. (2008), who reported that intermittent TM for 6 hours per day between E10 and E18 did not have a deleterious effect on embryo growth and chick weight. Taken together, these findings indicated that continuous 24 hours/day TM from E7 to E16 was too severe to support normal embryonic development, whereas intermittent 12 hours/day TM allowed embryos to dissipate excessive heat during the 12 hours of lower temperature and to avoid teratogenic consequences.

As for the effects during the post-hatching period, Christensen et al. (2000) demonstrated a long-lasting effect on body weight that started at hatching and persisted to marketing age. Indeed, Piestun et al. (2008) demonstrated a similar kind of effect, i.e. chicks with lower body weight at hatching (24H) did not exhibit compensatory growth sufficient to regain lost body weight before marketing age (Table 3). However, in both TM treatments a positive effect on broiler carcass quality characteristics was observed, namely, a significant increase in relative breast muscle weight and significant decrease in relative weight of the abdominal fat pad (Table 3).

Table 3. The effect of thermal manipulation (TM) during incubation on different parameters at hatching and at marketing age

Variables	Incubation Treatment[1]		
	Control	12 hours/day TM	24 hours/day TM
External piping without hatching (%)[2]	23[b]	33[b]	60[a]
Non-absorbed yolk and rough navel (%)[3]	5[c]	14[b]	34[a]
BW of males at hatching (g)	52.7[a]	52.9[a]	51.7[b]
BW of females at hatching (g)	52.4[a]	52.4[a]	51.4[b]
BW of males at 35 days of age (g)	2008[a]	2006[a]	1772[b]
BW of females at 35 days of age (g)	1756[a]	1722[a]	1600[b]
Male breast muscle (g/100 g BW)	18.20[b]	19.18[a]	19.30[a]
Female breast muscle (g/100 g BW)	18.06[b]	19.2[a]	18.85[a]
Male abdominal fat pad (g/100 g BW)	1.54[a]	1.17[b]	1.05[b]
Female abdominal fat pad (g/100 g BW)	1.99	1.91	1.88

[a,b,c] Within rows, means that possess different superscripts differ significantly ($P \leq 0.05$).

[1] Eggs were incubated under either control conditions (37.8°C and 56% RH) or thermal manipulated (TM) from E7 to E16 inclusive at 39.5°C and 65% RH continuously (24H treatment) or intermittently (12H treatment).

[2] External piping without hatching as a percentage of non-hatched chicks.

[3] As a percentage of hatched chicks.

The higher breast muscle relative weight suggested that TM during E7 through E16 promoted breast muscle growth. This was indicated by greater muscle hypertrophy from 2 weeks post-hatching through marketing age, which can be related to increased myofiber diameter that has been reported to require an increased number of muscle nuclei in the tissue (Allen et al., 1979; Cheek and Hill, 1970). Therefore, it was reasonable to hypothesize that TM caused an increased muscle cell proliferation in the embryo or post-hatching as was demonstrated by Piestun et al. (2009b) with another TM model. The effect on abdominal fat accumulation can be explained by the effect of TM on adipocyte development during embryogenesis. Hammond et al. (2007) found that TM of 38.5°C from E4 through E7 increased embryonic movement and associated energy expenditure, thereby reducing adipocyte diameter and size of the fat pad in the embryo. Since adipose tissue in the chick begins to form at around E12 (Speake et al., 1998), it can be speculated that TM, which partially coincided with embryonic adipose tissue formation, affected fat tissue formation.

Increased Temperature during Incubation - Acquisition of Thermotolerance

Thermal manipulations during incubation designed to improve thermotolerance have been based on the assumption that such manipulations must be conducted during a critical period that has a long lasting thermoregulatory effect based on epigenetic temperature adaptation. Epigenetic adaptation in poultry has been defined as a lifelong adaptation that occurs during embryogenesis or early post-hatching ontogeny, within critical developmental phases (Nichelmann and Tzschentke, 2002; Tzschentke and Basta, 2002; Tzschentke et al., 2004; Tzschentke and Plagemann, 2006). This definition has encompassed changes that regulate the expression of gene activity without altering the genetic structure of the primary DNA sequence. Molecular mechanisms of epigenetics, such as DNA methylation and chromatin remodeling, have been extensively reviewed (Bird, 2002; Bernstein et al., 2007; Kouzarides, 2007; Misteli, 2007; Schuettengruber et al. 2007; Zaratiegui et al., 2007).

Epigenetic temperature adaptation theory has been based on the assumption that environmental factors, especially ambient temperature (T_a), strongly affect determination of the 'set-point' for the temperature control system. During early development most functional systems have been reported to evolve from an open-loop system without a feedback mechanism into a closed control system with adequate feedback. Thus, TM during a critical phase of maturation of the feedback mechanism (usually in the post-hatching period), or during the development of the thermoregulatory axis (the hypothalamus–hypophysis–thyroid axis) during embryogenesis, might induce alterations in the thermoregulatory control system.

Broiler chicks complete development of their brain and thermoregulatory capacity by the age of 10 days post-hatching (Arad and Itsaki-Glucklish, 1991). Before this age body and brain temperatures have been shown to be maintained at lower levels than in adult chickens. Subsequently, with increasing age, the difference between body and brain temperatures increases linearly and significantly. The epigenetic response has been successfully modulated by TM of early post-hatching chicks by exploiting the incomplete maturation of the thermoregulatory system. For example, TM involving exposure of 3-day-old broiler chicks to 37-38°C at 60-80% relative humidity (RH) for 24 hours was found to improve acquisition of thermotolerance. The improvement achieved was manifested in the ability of the TM chicks to reduce heat production efficiently during exposure to acute thermal challenge at subsequent marketing age (Yahav and Hurwitz, 1996). This was accompanied by: (a) an alteration in sensible heat loss through convection and radiation (Yahav et al., 2005); (b) a significant reduction in the stress level of the TM chickens as indicated by their lower plasma corticosterone concentration; and (c) pronounced decrease in the 27-, 70- and 90-kDa heat-shock proteins (HSPs) in the heart muscle and lung tissue of TM chickens during thermal challenge (Yahav et al., 1997). It was suggested that induction of HSPs was correlated with body temperature (T_b), and that the HSP response was not part of the long-term mechanism elicited by the early-age TM. The reduction in heat production, coupled with increased sensible heat loss, enabled relatively slow development of hyperthermia and thus dramatically reduced mortality. The effect of TM on the preoptic/anterior hypothalamic region of the brain (PO/AH) of the chicks was studied with reference to two genes: R-Ras3 and the brain-derived neurotrophic factor (BDNF). The latter has been reported to initiate the internal cellular pathway of RAS that resulted in transduction of genes involved in neuronal growth and maintenance. In the chick PO/AH, significantly increased expression of R-Ras3 (Labunskay and Meiri, 2006), BDNF (Katz and Meiri, 2006), and 14-3-3ε (Meiri, 2008) were detected during TM, which suggested that these genes were involved in the thermoregulatory effect of the TM.

However, TM during embryogenesis, within a specific critical time period that corresponded to the formation and development of the energy-control axis, might be more efficient than TM post-hatching and might elicit a more pronounced effect on the thermoregulatory control system. Moreover, a commercial broiler house occupies a large area that encompasses a wide range of T_a, so that it would be difficult to achieve a uniform TM for 24 hours. Application of TM in the hatchery, where T_a and RH can be held within narrow ranges, was likely to overcome this obstacle.

As was mentioned previously, in contrast to the generally uniform temperature of commercial incubators, incubation conditions in nature have been found to be not so uniform, as a result of parental concerns involving searching for food and escaping from predators, and non-uniform nest insulation (Webb, 1987). This may be one of the reasons why birds in the wild have become quite able to cope with extreme environmental temperatures. Field measurements of incubation temperature have demonstrated that avian eggs usually experience temperature ranges of 30 to 40°C during their incubation (Webb, 1987), which may have led to an adequate capacity to cope with thermal challenge during the subsequent life span of the progeny. Exposing embryos to high or low temperatures during incubation, especially on certain days during the third trimester, improved their capacity to adapt to hot or cold environments, respectively, in the immediate post-hatching phase (Nichlmann et al., 1994; Tzschentke and Basta, 2002; Moraes et al., 2003; Yahav et al., 2004a).

The utilization of TM during incubation has to take three major parameters into consideration: the critical period, the temperature, and the duration of exposure. Determination of the critical period of embryogenesis that would enable successful application of TM to improve the acquisition of thermotolerance has been based on the hypothesis that the "set point" or "response threshold" of controlling systems could be altered most efficiently during development/maturation of the hypothalamus-hypophyseal-thyroid axis (thermoregulation) and/or the hypothalamus-hypophyseal-adrenal axis (stress). Until mid-incubation limited amounts of maternal thyroid hormones in the yolk have been reported (McNabb and King, 1993) and the thyroid gland has exhibited only limited ability to synthesize hormones. This period has been characterized by the synthesis of monoiodotyrosine on E7, of diiodotyrosine on E9, and of T_4 and thyroid-stimulating hormone (TSH) by the hypophysis on E10. The linkage of the hypothalamic–pituitary–thyroid axis has been shown to be formed only between E10.5 and E11.5 (Kameda et al., 1986). Levels of T_3 increased moderately beginning at E12, and there was a significant increase prior to hatching, probably because T_3 has been characterized as a permissive hormone that played a major role in final maturation (differentiation) of tissues and in the physiological integration of hatching. The hypothalamus-hypophyseal-adrenal axis developed from E10 through E15 inclusive. The two axes have been shown to be endocrinologically associated during incubation as corticotrophin releasing hormone (CRH) has been reported to stimulate thyroid stimulating hormone (TSH) excretion. The second parameter to be considered, temperature elevation, has to be significant to have a thermoregulatory effect, but a teratogenic effect could be developed as a result of an excessive incubation temperature (Romanoff, 1960). Such a deleterious effect might be directly related to the duration of exposure as well (the third parameter to be considered below): at any TM temperature, the longer the duration of the TM, the greater the possibility that a pronounced teratogenic effect might occur. Since there was a strong association between body weight and vitality of the hatched broiler chicks, on the one hand, and the productivity of the broiler during growth and at marketing, on the other hand (Noy and Sklan, 1999), this deleterious effect must be avoided.

In recent experiments (Piestun et al., 2008a, b) TM of 39.5°C and 65% RH was applied to broiler embryos between E7 and E16 (inclusive), either continuously (24H treatment) or intermittently for 12 h per day (12H treatment). These conditions were selected in light of three studies. The first was a search for the optimal critical period between E7 and E16 (inclusive); the intervals considered were: E7 to E10, E7 to E11, E7 to E12, E12 to E14, E12 to E15, E12 to E16, and E7 to E16 (inclusive of both days in each case) (Piestun et al., unpublished data). The second comprised previous studies concerning elevation of incubation temperature (Yahav et al., 2004a; Collin et al., 2005, 2007). The third study comprised a search for the optimal TM duration that had no negative effect on embryo development. Examined TM durations were 3, 6, 12, 15, 18, and 24 hours/day (Piestun, unpublished data). The results revealed a clear trend that the longer the TM duration, the better the effect on thermotolerance, but the more pronounced the deleterious effect on embryo development.

The acquisition of thermotolerance during embryogenesis by means of exposure to the above TM conditions was accompanied by significant reduction in embryonic metabolic rate during the endothermic phase of embryogenesis (the last days before hatching). This metabolic rate reduction was accompanied by reductions of eggshell temperature, heart rate, and oxygen consumption. In addition, the levels of plasma thyroid hormones, T_4 and T_3, were significantly lower in the TM-treated embryos than in the controls (Piestun et al., 2009a). The latter finding indicated a reduction in the threshold of hypothalamus-hypophysis-thyroid axis function with an important effect on thyroid gland activity (Piestun et al., 2009a).

In contrast to previous studies in which high temperatures were imposed during a similar critical period (Janke et al., 2002; Nichelmann and Tzschentke, 2002; Moraes et al., 2003; Nichelmann, 2004; Yahav et al., 2004a; Collin et al., 2007; Hulet, 2007; Leksrisompong et al., 2007; Lourens et al., 2007; Yalcin et al., 2008), the studies of Piestun et al. (2008a,b; 2009) found a long-lasting effect of TM applied during embryogenesis on the thermoregulatory abilities of the chicken during its life span. This effect was characterized by significantly lower T_b (0.2°C), which reduced heat production, and was coincident with significantly lower levels of plasma thyroid hormones. As a result, the TM-treated broilers exhibited lower energy demands for maintenance, which resulted in a commercially advantageous improved feed efficiency during the growth period, whether under normal or chronic hot conditions (Piestun et al., 2011). Moreover, chickens from both TM treatments exhibited an improved ability, as compared to controls, to cope with an acute heat challenge. Subjecting broiler chickens at 35 days of age to a thermal challenge of 35°C for 5 hours led to hyperthermia and mortality. Even so, the hyperthermia that developed in the TM chickens was significantly lower (by about 0.5°C) than that of the controls. Analysis of broiler T_b distribution at the end of the thermal challenge revealed a higher proportion of chickens with T_b above 45°C in the controls than among the TM broilers. The higher rate of hyperthermia in the controls coincided with a twofold greater mortality rate than that in the TM chickens. The improved ability of the TM broilers to regulate their T_b was attributed to the lower levels of plasma T_3 elicited during the thermal challenge, which indicated lower rates of heat production, on the one hand, and, on the other hand, greater capacity for sensible heat dissipation, mainly through convection and radiation, at the end of the thermal challenge. This emphasized the advantage of a better vasodilatation capacity. In addition, at the end of the thermal challenge plasma corticosterone was significantly lower in TM broilers than in the controls, indicating a lower level of stress during the heat exposure. This long-lasting beneficial effect of TM of the embryo on the thermoregulatory capabilities of the post-hatching broiler during its life span was probably a result of epigenetic temperature adaptation, which was characterized by an alteration of the energy balance axis set point / threshold response.

OVERALL CONCLUSION

- The evolutionary scheme has provided the hen with the potential to adaptively adjust the development of her progeny by exposing them to her hormonal and nutritional status in a manner that improved their efficiency to cope with predicted post-hatching environmental conditions. Such hormonal and nutritional exposure would certainly be influenced in the domestic fowl by management factors that influenced yolk sac absorption, which has become an issue central to "chick quality."
- There was no doubt that the interaction between egg storage period, storage temperature, storage position, and turning during incubation played a crucial role in the initiation of embryo development, mortality of embryos, the length of incubation, yolk sac absorption, hatchability, and chick quality.
- Pre-incubation of eggs reduced the "temperature shock" that resulted from the gradient of temperatures between storage and incubation and

- improved the uniformity of embryo development as well as hatchability and post-hatching development.
- Continuous TM from E7 to E16 was too severe to support normal embryonic development, whereas intermittent TM allowed embryos to dissipate excess heat and to avoid teratogenic consequences.
- The long-lasting positive effects of TM of the embryo on the thermoregulatory capabilities of the broiler during its life span was probably a result of epigenetic temperature adaptation, which was characterized by an alteration of the energy balance axis set point / threshold response.
- The development of the chick embryo was subject to maternal effects mediated by hormonal and nutritional deposition in the yolk. However, embryogenesis may also be affected by thermal manipulations, with long lasting effects post-hatching.

ACKNOWLEDGMENTS

Parts of the studies quoted in this chapter were supported by research grant No. IS-4198-09 from BARD, the United States – Israel Bi-national Agricultural Research and Development Fund and by research grant No. 356-0614 from the Egg and Poultry Board of Israel.

REFERENCES

Afşar, A., Elibol, O. and Brake, J. (2007). The effect of flock age and egg storage period on organ development and broiler performance. *Poult. Sci.,* 86 (Suppl. 1), 404.

Allen, R. E., Merkel, R. B. and Young, R. B. (1979). Cellular aspects of muscle growth: Myogenic cell proliferation. *J. Anim. Sci.,* 49, 115-127.

Alonso-Alvarez, C. (2006). Manipulation of primary sex ratio: an update review. *Avian Poult. Biol. Rev.,* 17, 1-20.

Arad, Z. and Itsaki-Glucklish, S. (1991). Ontogeny of brain temperature in quail chicks (*Coturnix coturnix japonica*). *Physiol. Zool.,* 64, 1356-1370.

Arora, K. L. and Kosin, I. L. (1966). Developmental responses of early turkey and chicken embryos to pre-incubation holding of eggs: Inter- and intra-species differences. *Poult. Sci.,* 45, 958-970.

Ates, C., Elibol, O. and Brake, J. (2004). The effect of storage period of eggs on hatching time and broiler performance. In: Proc. XXII World's Poult. Cong. Istanbul, Turkey; 4 pages (on CD).

Balthazart, J. and Ball, G. F. (1995). Sexual differentiation in brain and behavior in birds. *Trends Endocrinol. Metab.*, 6, 21-29.

Balthazart, J., Cornil, C. A., Charlier, T. D., Taziaux, M. and Ball, G.F. (2009). Estradiol, a key endocrine signal in the sexual differentiation and activation of reproductive behavior in quail. *J. Integr. Biol.*, 311, 323-345.

Becker, W. A. and Bearse, G. E. (1958). Pre-incubation warming and hatchability of chicken eggs. *Poult. Sci.*, 37, 944-948.

Becker, W. A., Spencer, J. V. and Hawkes, B. W. (1969). Angle of turning chicken eggs during storage. *Poult. Sci.*, 48, 1748 (abstract).

Benton, C. E. and Brake, J. (2001). Effects of presence of a blastoderm on albumen height and pH of broiler breeder eggs. *Poult. Sci.*, 80, 955-957.

Bernstein, B. E., Meissner, A. and Lander, E. S. (2007). The mammalian epigenome. Cell, 128, 669-681.

Biard, C., Surai, P. and Møller, A. P. (2005). Effects of carotenoid availability during laying on reproduction in the blue tit. *Oecologia* (Berlin), 144, 32-44.

Bird, A. (2002). DNA methylation patterns and epigenetic memory. Genes Dev., 16, 6-21.

Brake, J., Walsh, T. J., Benton, C. E., Petitte, J. N., Meijerhof, R. and Peñalva, G. (1997). Egg handling and storage. *Poult. Sci.*, 76, 144-151.

Brake, J., Plumstead, P. W., Lenfestey, B. A. and Elibol, O. (2004). Increased nutrient intake during rearing of broiler breeder pullets affects broiler progeny performance positively without effect on reproductive performance. In: Proc. XXII World's Poult. Congress: Istanbul, Turkey; 6 pages (on CD).

Brannan, K. E. M.S. Thesis: Effect of early incubation temperature and late incubation conditions on embryonic development and subsequent broiler performance. Raleigh, NC: The Graduate School, North Carolina State University; 2008.

Bull, J. J. (1980). Sex determination in reptiles. Q. Rev. Biol., 55, 3-21.

Carere, C. and Balthazart, J. (2007). Sexual versus individual differentiation: the controversial role of avian maternal hormones. *Trends Endocrinol. Metab.*, 18, 73-80.

Cheek, D. B. and Hill, D. E. (1970). Muscle and liver cell growth: Role of hormones and nutritional factors. *Fed. Proc.*, 29, 1503-1509.

Christensen, V. L., Donaldson, W. E. and Nestor, K. E. (1993). Effect of maternal dietary triiodothyronine on embryonic physiology of turkeys. *Poult. Sci.*, 72, 2316-2327.

Christensen, V. L., Donaldson, W. E., Nestor, K. E. and McMurtry, J. P. (1999). Effects of genetics and maternal dietary iodide supplementation on glycogen content of organs within embryonic turkeys. *Poult. Sci.*, 78, 890-898.

Christensen, V. L., Grimes, J. L., Donaldson, W. E. and Lerner, S. (2000). Correlation of body weight with hatchling blood glucose concentration and its relationship to embryonic survival. *Poult. Sci.*, 79, 1817-1822.

Christensen, V. L., Wineland, M. J., Fasenko, G. M. and Donaldson, W. E. (2001). Egg storage effects on plasma glucose and supply and demand tissue glycogen concentrations of broiler embryos. *Poult. Sci.,* 80, 1729-1735.

Collin, A., Picard, M. and Yahav, S. (2005). The effect of duration of thermal manipulation during broiler chick embryogenesis on body weight and body temperature of post-hatched chicks. *Anim. Res.*, 54, 105-111.

Collin, A., Berri, C., Tesseraud, S., Requena, F., Cassy, S., Crochet, S., Duclos, M. J., Rideau, N., Tona, K., Buyse, J., Bruggemann, V., Decuypere, E., Picard, M. and Yahav, S. (2007). Effects of thermal manipulation during early and late embryogenesis on thermotolerance and breast muscle characteristics in broiler chickens. *Poult. Sci.*, 86, 795-800.

Decuypere, E. and Michels, H. (1992). Incubation temperature as a management tool: A review. *World's Poult. Sci. J.,* 48, 28-38.

Druyan, S. (2010). The effects of genetic line (broilers vs. layers) on embryo development. *Poult. Sci.*, 89, 1457-1467.

Druyan, S., Levi, E., Shinder, D. and Stern, T. (2012). Reduced O_2 concentration during the CAM development - its effect on physiological parameters of broilers embryos. *Poult Sci.*, 91, 987-997.

Ebensperger, C., Drew, U., Mayerova, A. and Wolf, U. (1988). Serological H-Y antigen in the female chicken occurs during gonadal differentiation. *Differentiation*, 37, 186-191.

Eggert, C. (2004). Sex determination: the amphibian models. *Reprod. Nutr. Dev.*, 44, 539-549.

Eiby, Y. A., Wilmer, J. W. and Booth, D. T. (2008). Temperature-dependent sex-biased embryo mortality in a bird. *Proc. Royal Soc. B*, 275, 2703-2706.

Eising, C. M., Müller, W. and Groothuis, T. G. G. (2006). Avian mothers create different phenotypes by hormone deposition in their eggs. *Biol. Lett.*, 2, 20-22.

Eising, C. M., Visser, G. H., Müller, W. and Groothuis, T. G. G. (2003). Steroids for free? No metabolic costs of elevated maternal androgen levels in the black-headed gull. *J. Exp. Biol.*, 206, 3211-3218.

Elf, P. K. and Fivizzani, A. J. (2002). Changes in sex steroid levels in yolks of the leghorn chicken *Gallus domesticus* during embryonic development. *J. Exp. Zool.*, 293, 594-600.

Elibol, O., and Brake, J. (2008). Effect of egg position during 3 and 14 days of storage and turning frequency during subsequent incubation on hatchability of broiler hatching eggs. *Poult. Sci.*, 87, 1237-1241.

Elibol, O., Peak, S. D. and Brake, J. (2002). Effect of flock age, length of egg storage, and frequency of turning during storage on hatchability of broiler hatching eggs. *Poult. Sci.*, 81, 945-950.

Eusebio-Balcazar, P. E. M. S. Thesis: Effect of breeder nutrition and feeding program during rearing and production on broiler leg health. Raleigh, NC: The Graduate School, North Carolina State University; 2009.

Ewert, M. A. and Nelson, C. E. (1991). Sex determination in turtles: diverse patterns and some possible adaptive values. *Copia*, 4, 50-59.

Eyal-Giladi, H. and Kochav, S. (1976). From cleavage to primitive streak formation: a complementary table and a new look at first stages of the development of the chick. I. General morphology. *Dev. Biol.*, 49, 321-337.

Fasenko, G. M. (2007). Egg storage and the embryo. *Poult. Sci.*, 86, 1020-1024.

Fasenko, G. M., Robinson, F. E., Hardin, R. T. and Wilson, J. L. (1992). Variability in pre-incubation embryonic development in domestic fowl. 2. Effect of duration of egg storage period. *Poult. Sci.,* 71, 2129-2132.

Fasenko, G. M., Robinson, F. E., Segura, J. C., Feddes, J. J. R. and Ouellette, C. A. (2002). Long term hatching egg storage alters the metabolism of broiler embryos. *Poult. Sci.*, 80 (Suppl. 1), 62.

Ferguson, M. W. J. (1994). Temperature depended sex determination in reptiles and manipulation of poultry sex by incubation temperature. In: Proc. 9th European Poult. Conf.: Glasgow, Scotland; pp. 380-382.

Funk, E. M., Forward, J. and Kempster, H. C. (1960). Effect of holding temperature on hatchability of eggs. *Missouri Agr. Exp. Sta. Bull.*, 539. Columbia. MO.

Funk, E. M. and Forward, J. (1960). Effect of holding temperature on hatchability of chicken eggs. *Missouri Agr. Exp. Sta. Bull.*, 732. Columbia. MO.

Gasparini, J., McCoy, K. D., Haussy, C., Tveraa, T., and Boulinier, T. (2001). Induced maternal response to the Lyme disease spirochete *Borrelia burgdorferi* sensu lato in a colonial seabird, the kittiwake *Rissa tridactyla*. *Proc. Royal Soc. B Biol. Sci.*, 268, 647-650.

Gil, D. Physiology, Endocrinology and Behavior, Advances in the study of behavior: Hormones in avian eggs. Vol. 38. San Diego: Elsevier Academic Press Inc.; 2008; pp. 337-398.

Gilbert, L., Bulmer, E., Arnold, K. E. and Graves, J. A. (2007). Yolk androgens and embryo sex: Maternal effects or confounding factors? *Hormones and Behav.*, 51, 231-238.

Goodrum, J. W., Britton, W. M. and Davis, J. B. (1989). Effect of storage conditions on albumen pH and subsequent hard-cooked eggs peelability and albumen shear strength. *Poult. Sci.*, 68, 1226-1231.

Göth, A. (2007). Incubation temperature and sex ratios in Australian brush-turkey (*Alectura lathami*) mounds. *Austral. Ecol.*, 32, 378-385.

Göth, A. and Booth, D. T. (2005). Temperature dependent sex ratio in a bird. *Biol. Lett.*, 1, 31-33.

Groothuis, T. G. G., Müller, W., von Engelhardt, N., Carere, C. and Eising, C. (2005). Maternal hormones as a tool to adjust offspring phenotype in avian species. *Neurosci. Behav. Rev.*, 29, 329-352.

Groothuis, T. G. G. and Swabl, H. (2002). The influence of laying sequence and habitat characteristics on maternal yolk hormone levels. *Func. Ecol.*, 16, 281-289.

Groothuis, T. G. G. and Swabl, H. (2008). Hormone-mediated maternal effects in birds. In: Mechanisms matter but what do we know about them? *Phil. Trans. R. Soc. B,* 363, 1647-1661.

Güçbilmez, M., Elibol, O. and Brake, J. (2009). Effect of rate of preincubation temperature increase on hatchability of broiler hatching eggs. Poult. Sci. Abstracts, p. 47 on CD and at www.poultryscience.org.

Güçbilmez, M., Özlü, S., Shiranjang, R., Elibol, O. and J. Brake, J. (2010). Effects of heating broiler hatching eggs during 6 or 11 days of storage on hatchability. Poult. Sci. 89, E-Suppl. 1, 775 on CD and at www.poultryscience.org.

Hammond, C. L., Simbi, B. H. and Stickland, N. C. (2007). In ovo temperature manipulation influences embryonic motility and growth of limb tissues in the chick (*Gallus gallus*). *J. Exp. Biol.*, 210, 2667-2675.

Havenstein, G. B., Ferket, P. R., Scheideler, S. E. and Larson, B. T. (1994a). Growth, livability, and feed conversion of 1957 vs 1991 broilers when fed "typical" 1957 and 1991 broiler diets. *Poult. Sci.*, 73, 1785-1794.

Havenstein, G. B., Ferket, P. R., Scheideler, S. E. and Rives, D. E. (1994b). Carcass composition and yield of 1991 vs. 1957 broilers when fed "typical" 1957 and 1991 broiler diets. *Poult. Sci.*, 73, 1795-1804.

Havenstein, G. B., Ferket, P. R. and Qureshi, M. A. (2003a). Growth, livability and feed conversion of 1957 versus 2001 broilers when fed representative 1957 and 2001 broiler diets. *Poult. Sci.*, 82, 1500-1508.

Havenstein, G. B., Ferket, P. R. and Qureshi, M. A. (2003b). Carcass composition and yield of 1957 vs. 2001 broilers when fed representative 1957 and 2001 broiler diets. *Poult. Sci.*, 82, 1509-1518.

Hays, F. A. and C. Nicolaides. (1934). Variability in development of fresh-laid hen eggs. *Poult. Sci.*, 13, 74-80.

Hayward, L. S. and Wingfield, J. C. (2004). Maternal corticosterone is transferred to avian yolk and may alter offspring growth and adult phenotype. *Gen. Comp. Endocrinol.* 135, 365-371.

Hill, D. (2001). Chick length uniformity profiles as a field measurement of chick quality? *Avian Poult. Biol. Rev.*, 12, 188.

Hulet, R. M. (2007). Managing incubation: Where are we and why? *Poult. Sci.*, 86, 1017-1019.

Insko, Jr., W. M. The Fertility and Hatchability of Chicken and Turkey Eggs: Physical conditions in incubation. In: L.W. Taylor, editor. London: John Wiley and Sons, Inc.; 1949; pp. 210-243.

Isakson, C., Magrath, M. J. L., Groothuis, T. G. G. and Komdeur, J. (2010). Androgens during development in a bird species with extremely sexually dimorphic growth, the brown songlark, *Cinclorhamphus cruralis*. *Gen. Comp. Endocrinol.*, 165, 97-103.

Janke, O., Tzschentke, B. and Boerjan, B. (2004). Comparative investigations of heat production and body temperature in modern chicken breeds. *Avian Poult. Biol. Rev.*, 15, 191-196.

Kameda, Y., Udatsu, K., Horino, M. and Tagawa, T. (1986). Localization and development of immunoreactive triiodothyronine in thyroid glands of dogs and chickens. *Anat. Rec.*, 214, 168-176.

Katz, A. and Meiri, N. (2006). Brain-derived neurotrophic factor is critically involved in thermal-experience-dependent developmental plasticity. *J. Neurosci.* 26, 3899-3907.

Kirk, S., Emmans, G. C., McDonald, R. and Arnot, D. (1980). Factors affecting the hatchability of eggs from broiler breeders. *Br. Poult. Sci.*, 21, 37-53.

Kosin, I. L. and St. Pierre, E. (1956). Studies on pre-incubation warming of chicken and turkey eggs. *Poult. Sci.*, 35, 1384-1392.

Kouzarides, T. (2007). Chromatin modifications and their function. *Cell*, 128, 693-705.

Labunsky, G. and Meiri, N. (2006). R-Ras3/(M-Ras) is involved in thermal adaptation in the critical period of thermal control establishment. *J. Neurobiol.*, 66, 56-70.

Landauer, W. (1967). The hatchability of chicken eggs as influenced by environment and heredity. *Storrs Agric. Exp. Stat. Mono.* 1 (revised). Storrs, CT.

Leksrisompong, N. Ph.D. Dissertation: Effects of feeder space and feeding programs during rearing and laying on broiler breeder reproductive performance and livability. Raleigh, NC: The Graduate School, North Carolina State University; 2010.

Leksrisompong, N., Romero-Sanchez, H., Plumstead, P. W., Brannan, K. E. and Brake, J. (2007). Broiler incubation. 1. Effect of elevated temperature during late incubation on body weight and organs of chicks. *Poult. Sci.*, 86, 2685-2691.

Lephart, E. D. (1996). A review of brain aromatase cytochrome P450. *Brain Res. Rev.* 22, 1-26.

Lin, Y. M., Yahav, S., Elibol, O. and Brake, J. (2010). Effects of turning frequency during incubation on broiler embryonic development. Poult. Sci. 89, E-Suppl. 1, 222 on CD and at www.poultryscience.org.

Lin, Y-M. M. S. Thesis: Effect of preheating temperature, early incubation temperature, and different turning frequency on embryonic development and broiler performance. Raleigh, NC: The Graduate School, North Carolina State University; 2011.

Love, O. P., Chin, E. H., Wynne-Edwards, K. E. and Williams, T. D. (2005). Stress hormones: a link between maternal condition and sex biased reproductive investment. *Am. Nat.*, 166, 751-766.

Lourens, A., Van den Brand, H., Heetkamp, M. J., Meijerhof, R. and Kemp, B. (2007). Effects of eggshell temperature and oxygen concentration on embryo growth and metabolism during incubation. *Poult. Sci.*, 86, 2194-2199.

Lundy, H. The Fertility and Hatchability of the Hen's Egg: A review of the effects of temperature, humidity and gaseous exchange environment in the incubator on the hatchability of the hen's eggs. In: T. C. Carter and B. M. Freeman, editors. Edinburgh, UK: Oliver and Boyd; 1969; pp.143-176.

Mather, C. M. and Laughlin, K. F. (1977). Storage of hatching eggs: The effect on early embryonic development. *Br. Poult. Sci.*, 18, 597-603.

Mayes, F. J. and Takeballi, M. A. (1984). Storage of the eggs of the fowl (*Gallus domesticus*) before incubation: a review. *World's Poult. Sci. J.*, 40, 131-140.

McNabb, F. M. A. and King, D. B. The endocrinology of growth, development, and metabolism of vertebrates: Thyroid hormone effects on growth, development and metabolism. In: M. P. S. Schreibman and P. K. T. Pang, editors. New York: Academic Press; 1993; pp. 393–417.

Meijerhof, R. (1992). Pre-incubation holding of hatching eggs. *World's Poult. Sci. J.* 48, 57-68.

Misteli, T. (2007). Beyond the sequence: cellular organization of genome function. *Cell*, 128, 787-800.

Moraes, V. M. B., Malehiros, D., Bruggemann, V., Collin, A., Tona, K., Van As, P., Onagbesan, O. M., Buyse, J., Decuypere, E., and Macari, M. (2003). Effect of thermal conditioning during embryonic development on aspects of physiological responses of broilers to heat stress. *J. Therm. Biol.*, 28, 133-140.

Mousseau, T. A. and Fox, C. W. Maternal effects. In: T. A. Mousseau and C. W. Fox, editors. New York: Oxford University Press; 1998.

Müller, W., Goerlich, W. C., Vergauwen, J., Groothuis, T. G. G. and Eens, M. (2012). Sources of variations in yolk hormone deposition: Consistency, inheritance and developmental effects. *Gen. Comp. Endocrinol.*, 175, 337-343.

Nelson, R. J. An introduction to behavioral endocrinology. MA: Sinauer, Sunderlans; 2000.

Nichelmann, M., Lange, B., Pirow, R., Langbein, J. and Herrmann, S. Avian Thermal Balance in Health and Disease: Thermoregulation during the perinatal period. In: Advances in Pharmacological Science. E. Zeisberger, E. Schönbaum, and P. Lomax, editors. Basel, Switzerland: Birkhäuser Verlag; 1994; pp.167–173.

Nichelmann, M. and Tzschentke, B. (2002). Ontogeny of thermoregulation in precocial birds. *Comp. Biochem. Physiol. A Mol. Integr. Physiol.*, 131, 751-763.

Nishikimi, N., Kanasaku, N., Saito, N., Usami, M., Ohno, Y. and Shimada, K. (2000). Sex differentiation and mRNA expression of P450c17, P450arom and AMH in gonads of the chicken. *Mol. Reprod. Dev.*, 55, 20-30.

Noy, Y. and Sklan, D. (1999). Energy utilization in newly hatched chicks. *Poult. Sci.*, 78, 1756-1750.

Oldfield, R. G. (2005). Genetic abiotic and social influences on sex differentiation in cichlid fishes and the evolution of sequential hermaphrodism. *Fish.* 6, 93-110.

Peebles, E. D., Gardner, C. W., Brake, J., Benton, C. E., Bruzual, J. J. and Gerard, P. D. (2000). Albumen height and yolk and embryo compositions in broiler hatching eggs during incubation. *Poult. Sci.*, 79, 1373-1377.

Pieau, C., Dorizzi, M. and Richard-Mercier, N. (1999). Temperature-dependent sex determination and gonadal differentiation in reptiles. *Cell Mol. Life Sci.*, 55, 887-900.

Piestun, Y., Shinder, D., Ruzal, M., Halevy, O., Brake, J. and Yahav, S. (2008a). Thermal manipulations during broiler embryogenesis: effect on the acquisition of thermotolerance. *Poult. Sci.*, 87, 1516-1525.

Piestun, Y., Shinder, D., Ruzal, M., Halevy, O. and Yahav, S. (2008b). The effect of thermal manipulations during the development of the thyroid and adrenal axes on in-hatch and post-hatch thermoregulation. *J. Therm. Biol.*, 33, 413-418.

Piestun, Y., Halevy, O. and Yahav, S. (2009a). Thermal manipulations of broiler embryos - the effect on thermoregulation and development during embryogenesis. *Poult. Sci.*, 88, 2677-2688.

Piestun, Y., Harel, M., Barak, M., Yahav, S. and Halevy, O. (2009b). Thermal manipulations in late-term chick embryos have immediate and longer term effects on myoblast proliferation and skeletal muscle hypertrophy. *J. Appl. Physiol.* 109(1), 233-240.

Piestun, Y., Halevy, O., Shinder, D., Ruzal, M., Druyan, S. and Yahav, S. (2011). Thermal manipulations during broiler embryogenesis improves post hatch performance under hot conditions. *J. Therm. Biol.* 36, 469-474.

Piestun, Y., Druyan, S., Brake, J. and Yahav, S. (2012). Thermal treatments prior to and during the beginning of incubation affect phenotypic characteristics of broiler chickens post hatching. *Poult. Sci.*, (submitted).

Piferrer, F., Balzquez, M., Navarro, L. and Gonzalez, A. (2005). Genetic, endocrine and environmental components of sex determination and differentiation in the European sea bass. *Gen. Comp. Endocrinol.* 142, 102-110.

Proudfoot, F. G. and Hulan, H. W. Agriculture Canada Publications 1573/E, Research Station: Care of hatching eggs before incubation. Kentville, NS, Canada: 1976; pp.1-17.

Quinn, A. E., Georges, A., Sarre, S. D., Guarino, F., Ezaz, T. and Graves, J. A. M. (2007). Temperature sex reversal implies sex gene dosage in a reptile. *Science*, 316, 411.

Räsanen, K. and Kruuk, L. E. B. (2007). Maternal effects and evolution at ecological time scales. *Func. Ecol.,* 21, 408-421.

Reijrink, I. A. M. Ph.D. Thesis: Storage of hatching eggs - Effects of storage and early incubation conditions on egg characteristics, embryonic development, hatchability, and chick quality. Wageningen University, the Netherlands; 2010.

Renema, R. A., Feddes, J. J. R., Schmid, K. L., Ford, M. A. and Kolk, A. R. (2006). Internal egg temperature in response to preincubation warming in broiler breeder and turkey eggs. *J. Appl. Poult. Res.* 15, 1-8.

Ricklefs, R. E. (1987). A comparative analysis of avian embryonic growth. *J. Exp. Zool.* 51, 309-324.

Roff, D.A. Maternal effects as adaptations: Detection and measurement of maternal effects. In: T. A. Mousseau and C. W. Fox, editors. New York: Oxford University Press, 1998.

Romanoff, A. L. The Avian Embryo. New York: J. Wiley and Sons, Inc.; 1960.

Romero-Sanchez, H., Plumstead, P. W., Leksrisompong, N., Brannan, K. E. and Brake, J. (2007). Effect of plane of rearing nutrition of broiler breeder females on body weight, egg production, fertility, and progeny performance. *Poult. Sci.,* 86, 1553.

Rubolini, D., Romano, M., Martinelli, R., Leoni, B. and Saino, N. (2006). Effects of prenatal yolk androgens on armaments and ornaments of the ring-necked pheasant. *Behav. Ecol. Sociobiol.*, 59, 549-560.

Rudnick, D. (1944). Early history and mechanics of the chick blastoderm: A Review. *Q. Rev. Biol.*, 19, 187-212.

Russell, A. F. and Lummaa, V. (2009). Maternal effects in cooperative breeders: from hymenopterans to humans. *Phil. Trans. Royal Soc., B* 364, 1143-1167.

Rutkowska, J. and Badyaev, A. V. (2008). Meiotic drive and sex determination: molecular and cytological mechanisms of sex ratio adjustment in birds. *Phil. Trans. Royal Soc. Lond. B. Biol. Sci.*, 363, 1675-1686.

Rutkowska, J. and Cichon, M. (2006). Maternal testosterone affects the primary sex ratio and offspring survival in zebra finches. *Animal. Behav.*, 71, 1283-1288.

Ryan, B. C. and Vandenbergh, J. G. (2002). Intrauterine position effects. *Neurosci. Behav. Rev.*, 26, 133-151.

Sakata, N., Tamori, Y. and Wakahara, M. (2005). P450 aromatase expression in the temperature-sensitive sexual differentiation of salamander (*Hynobius retardatus*) gonads. *Int. J. Dev. Biol.*, 49, 417-425.

Sandercock, D. A., Mitchell, M. A. and MacLeod, M. G. (1995). Metabolic heat production in fast and slow growing broiler chickens during acute heat stress. *Br. Poult. Sci.*, 36, 868-869.

Scheib, D. (1983). Effects and role of estrogens in avian gonadal differentiation. *Differentiation*, 23, S87-S92.

Schuettengruber, B., Chourrout, D., Vervoort, M., Leblanc, B. and Cavalli, G. (2007). Genome regulation by polycomb and trithorax proteins. *Cell*, 128, 735-745.

Schwabl, H. (1993). Yolk is source of maternal testosterone for developing birds. *Proc. Natl. Acad. Sci.* 90, 11446-11450.

Shimada, K. (1998). Gene expression of steroidogenic enzymes in chicken embryonic gonads. *J. Exp. Zool.*, 281, 450-456.

Shine, R., Elphick, M. J. and Donnellan, S. (2002). Co-occurrence of multiple, supposedly incompatible modes of sex determination in lizard population. *Ecol. Lett.*, 5, 486-489.

Siano, N., Ferrari, R. P., Romano, M., Martinelli, R. and Møller, A. P. (2003). Experimental manipulation of egg carotenoids affects immunity of barn swallow nestlings. *Proc. Royal Soc. B. Biol. Sci.*, 270, 2485-2489.

Smith, C. A., Roeszler, K. N., Hudson, Q. J. and Sinclair, A. H. (2007). Avian sex determination: what, when, and where? *Cytogen. Genome Res.*, 117, 165-173.

Smith, C. A. and Sinclair, A. H. 2004. Sex determination: insights from the chicken. *Bioassays,* 26, 1-13.

Sockman, K. W. and Schwabl, H. (2000). Yolk androgens reduce offspring survival. *Proc. Royal Soc. Lond., B* 267, 1451-1456.

Speake, B., Murray, A. and Noble, R. (1998). Transport and transformations of yolk lipids during development of the avian embryo. *Prog. Lipid Res.*, 37, 1-32.

Spotila, J. R., Weinheimer, C. J. and Paganelli, C. V. (1981). Shell resistance and evaporative water loss from bird eggs: Effect of wind speed and egg size. *Physiol. Zool.*, 54, 195-202.

Stern, C. D. Avian Incubation: The sub-embryonic fluid of the egg of the domestic fowl and its relationship to the early development of the embryo. In: S.G. Tullett, editor. London: Butterworth-Heinemann; 1991; pp. 81-90.

Taylor, L. W. and C. A. Gunns. (1939). Development of the unincubated chick embryo in relation to hatchability of the egg. In: Proc. 7th World's Poult. Cong., pp. 188-190.

Tzschentke, B. and Basta, D. (2002). Early development of neuronal hypothalamic sensitivity in birds: influence of epigenetic temperature adaptation. *Comp. Biochem. Physiol. A,* 131, 825-832.

Tzschentke, B., Basta, D., Janke, O. and Maier, I. (2004). Characteristics of early development of body functions and epigenetic adaptation to the environment in poultry: focused on development of central nervous mechanisms. *Avian Poult. Biol. Rev.,* 15, 107-118.

Tzschentke, B. and Plagemann, A. (2006). Imprinting and critical periods in early development. *World's Poult. Sci. J.,* 62, 626-638.

Veiga, J. P., Vinuela, J., Cordero, P. J., Aparicio, J. M. and Polo, V. (2004). Experimentally increased testosterone affects social rank and primary sex ratio in spotless starling. *Horm. Behav.,* 46, 47-53.

Walsh, T. J., Rizk, R. E. and Brake, J. (1995). Effects of storage for 7 or 14 days at two temperatures in the presence and the absence of carbon dioxide on albumen characteristics, weight loss and early embryonic mortality of broiler hatching eggs. *Poult. Sci.,* 74, 1403-1410.

Webb, D. R. (1987). Thermal tolerance of avian embryos: A review. Condor, 89, 874-898.

Wilson, H. R. (1991). Inter-relationships of egg size, chick size, posthatching growth and hatchability. *World's Poult. Sci. J.,* 47, 5-20.

Yahav, S. (2009). Alleviating heat stress in domestic fowl - different strategies. *World's Poult. Sci. J.,* 65, 719–732.

Yahav, S., Collin, A., Shinder, D., and Picard, M. (2004a). Thermal manipulations during broiler chick's embryogenesis - the effect of timing and temperature. *Poult. Sci.,* 83, 1959-1963.

Yahav, S. and Hurwitz, S. (1996). Induction of thermotolerance in male broiler chickens by temperature conditioning at an early age. *Poult. Sci.,* 75, 402-406.

Yahav, S., Shamai, A., Horev, G., Bar-Ilan, D., Genina, O. and Friedman-Einat, M. (1997). Effect of acquisition of improved thermotolerance on the induction of heat shock proteins in broiler chickens. *Poult. Sci.,* 76, 1428-1434.

Yahav, S., Shinder, D., Ruzal, M., Gilo, M. and Piestun, Y. Body temperature control: Controlling body temperature - the opportunities for highly productive domestic fowl. In: A. B. Cisneros and. B. L. Goins, editors. New York, NY: NovaScience Publishers; 2009; pp. 65-98.

ılcin, S., Cabuk, M., Bruggeman, V., Babacanoglu, E., Buyse, J. Decuypere, E. and Siegel, P. B. (2008). Acclimation to heat during incubation. 1. Embryonic morphological traits, blood biochemistry, and hatching performance. *Poult. Sci.,* 87, 1219-1228.

oshida, K., Shimada, K. and Saito, N. (1996). Expression of p45017alfa hydroxylase and P450 aromatase genes in the chicken gonad before and after sexual differentiation. *Gen. Comp. Endocrinol.,* 102, 233-240.

Zhao, D., McBride, D., McQueen, H. A., McGrew, M. J., Hocking, P. M., Lewis, P. D., Sang, H. M. and Clinton, M. (2010). Somatic sex identity is cell autonomous in the chicken. *Nature,* 464, 237-243.

Zaratiegui, M., Irvine, D. V. and Martienssen, R. A. (2007). Noncoding RNAs and gene silencing. *Cell*, 128, 763-776.

In: Embryo Development
Editors: D. Reyes and A. Casales

ISBN: 978-1-62417-723-1
© 2013 Nova Science Publishers, Inc.

Chapter 2

MOLECULAR AND CELLULAR ASPECTS OF BLASTOCYST DORMANCY AND REACTIVATION FOR IMPLANTATION

Zheng Fu[1,2,*], Yongjie Chen[1,2,*], Weiwei Wu[1,2], Shumin Wang[1], Weixiang Wang[1], Bingyan Wang[1] and Haibin Wang[1,†]

[1]State Key Laboratory of Reproductive Biology, Institute of Zoology, Chinese Academy of Sciences, Beijing, PR China
[2]Graduate School of the Chinese Academy of Sciences, Beijing, PR China

ABSTRACT

Blastocyst activation, a process for the blastocyst to achieve implantation competency is equally important as attainment of uterine receptivity for the success of embryo implantation. While a wide range of regulatory molecules have been identified as essential players in conferring uterine receptivity in both laboratory animal models and humans, it remains largely unknown how blastocysts achieve

---

[*] These authors contribute equally.
[†] Correspondence: Haibin Wang, PhD, State Key Laboratory of Reproductive Biology, Institute of Zoology, Chinese Academy of Sciences, 1 Beichen West Road, Chaoyang District, Beijing 100101, PR China; E-mail: hbwang@ioz.ac.cn; Tel: +86-10-64807868; Fax: +86-10-64807099.

implantation competency. This chapter will highlight our current knowledge about the mechanisms governing the process of blastocyst activation. A better understanding of this periimplantation event is hoped to alleviate female infertility and help to develop novel contraceptives and new strategies for accessing embryo quality in clinical practice.

Keywords: Embryonic diapause, blastocyst activation, implantation

1. INTRODUCTION

Formation of a blastocyst with two distinct cell lineages, the outer specialized trophectodermal epithelium and the inner cell mass, is the climax event of early embryo development prior to implantation (Cockburn and Rossant, 2010, Wang and Dey, 2006). In mammals, early embryos only at the blastocyst stage can implant into the uterine womb. However, in nearly 100 species, development of preimplantation embryos can be suspended at the blastocyst stage without further initiation of attachment reaction for some time periods (Lopes, et al., 2004, Mead, 1993, Renfree and Shaw, 2000). This phenomenon is known as delayed implantation or embryonic diapause. Under this physiological condition, the blastocyst enters a stage, namely blastocyst dormancy, while the uterus enters a neutral status. However, dormant blastocysts can be reactivated at certain circumstances and reinitiate attachment with uterine epithelium for further development.

It has been generally accepted that successful implantation requires a synchronization of the blastocyst to achieve implantation competency and the uterus to enter into a receptive state that accepts and accommodates the implanting blastocyst (Dey, et al., 2004, Lim and Wang, 2010, Tranguch, et al., 2005, Wang and Dey, 2006). This embryo-uterine attachment reaction involving intimate two-way interactions can only occur in a self-limited short period, namely the "window of implantation" in mice, humans and other species (Ma, et al., 2003, Paria, et al., 1993b, Rogers and Murphy, 1989, Yoshinaga, 1980). Increasing evidence shows that failure to achieve 'on-time' implantation adversely affects the pregnancy outcome: a short delay in the attachment of embryos to the wall of the uterus during early pregnancy will create series adverse ripple effects on later developmental processes (Song, et al., 2002, Ye, et al., 2005). In humans, evidence also shows that implantation beyond the normal window leads to spontaneous pregnancy losses (Wilcox, et al., 1999), further highlighting the emerging concept that quality of implantation determines the quality of term pregnancy (Dey, 2005).

With the advance of gene expression studies and the development of genetically engineered mouse models, the molecular and cellular events that confer uterine receptivity have been extensively explored with identification of a wide range of regulatory molecules (Dey, et al., 2004, Lim and Wang, 2010, Tranguch, et al., 2005, Wang and Dey, 2006). However, it remains largely unknown as to the underlying mechanism, by which blastocysts achieve implantation competency. It is of particular interests to know how normal blastocysts undergo dormancy, survive for an extended period and then resume activation for implantation. In this chapter, we summarize the historical aspect and recent progress on the potential mechanism governing blastocyst dormancy and activation in various model systems, primarily focusing on molecular and cellular aspects of blastocyst functions during embryonic diapause and reactivation. The knowledge is hoped to facilitate the development of new strategies to correct implantation failure and improvement of pregnancy rates in women.

2. EMBRYONIC DIAPAUSE AND REACTIVATION: A UNIQUE STRATEGY FOR IMPLANTATION AT OPTIMAL CIRCUMSTANCES

Embryonic diapause characterized by delayed implantation occurs in nearly 100 mammals (about 70 eutherian species and 30 marsupial species) (Renfree and Shaw, 2000). The earliest records for delayed implantation can be traced back to the 19th century (Franklin, 1958). In 1854, Bishchoff reported a delayed implantation phenomenon in roe deer. In 1880, Fries recorded a delayed implantation in the European badger. In 1891, Lataste discussed five species of rodents in which delayed implantation occurred (Franklin, 1958). Daniel and King observed a lengthened gestation period in white mice and white rats sucking young during pregnancy, pointing toward to a lactation-induced delayed implantation (Daniel, 1910, King, 1913).

This evolutionarily conserved phenomenon has been classified into two functionally distinct categories - the obligate diapause and the facultative diapause (Table 1). Obligate diapause also known as seasonal delayed implantation existing in every gestation of a species, is believed to be a mechanism for synchronizing parturition with favorable environmental conditions for neonatal survival; while facultative diapause, well-known in rodents and marsupials, happens on environmental conditions related to survival of dam or suck lactation (Lopes, et al., 2004). The length of embryonic diapause ranges from several days to years in various species. Embryonic diapause can be induced, maintained and terminated by many environmental stimuli, such as length of photoperiod, metabolic stress (CK, 1940), lactation (McLaren, 1969), social stress (Marois, 1982) and artificial controlling of hormone steroids (Yoshinaga and Adams, 1966). This phenomenon appears most commonly at the blastocyst stage. The blastocysts cease its cell mitotic activity or undergo a short period of slow growth during diapause, while resumption of blastocyst cellular activities and development happen soon after endocrine termination of embryonic diapause. In bats, reproductive delays are achieved through several mechanisms, such as delayed fertilization, delayed implantation, and delayed development (Renfree and Shaw, 2000). Delayed development may also occur in Australian sea lions (Gales, et al., 1997). However, delayed implantation does not occur in certain species including hamsters, rabbits, guinea pigs or pigs. Whether this phenomenon occurs in humans or not is not known.

Endocrine regulation of embryonic diapause has been well established in many species (Table1). As a critical endocrine regulator of embryonic diapause, pituitary hormone prolactin can either support or inhibit ovarian corpus luteal function, thus to initiate or terminate diapause in different species (Figure 1). For example, prolactin is necessary for blastocyst and luteal development in mustelid carnivores, such as the mink and spotted skunk (Douglas, et al., 1997, Murphy, et al., 1993a, Polejaeva, et al., 1997). Its circulating concentrations increase some days prior to implantation (Mead, 1993, Murphy and Rajkumar, 1985). Prolactin alone can terminate embryonic diapause and induce implantation in hypophysectomized mink (Murphy, et al., 1981), while inhibition of prolactin secretion with dopamine agonists prevents implantation (Murphy, 1983, Papke, et al., 1980). For the majority of seasonal breeding mammals prolactin secretion is mainly regulated by photoperiod through regulation of pineal gland secretion of melatonin. In mink, facultative day length or the time exposure to light induce or terminate diapause (Murphy and James, 1974). Chronic melatonin treatment via inhibiting the secretion of prolactin prevents the termination of diapause, while supplement of exogenous prolactin reinitiates implantation in mink (Murphy, et al., 1990).

Table 1. Embryonic diapause in mammals

	Characters	**Hormone regulation**	**Representative animals**	**Blastocyst state**
Facultative diapause	Induced by environmental conditions related to survival or dam, such as lactation, metabolic stress, or experimental delayed model in rodents	Prolactin secretion during lactation Low estrogen levels	Mice and rats	No further growth in diapause status, blastocysts without zona pellucida
		Prolactin secretion for diapause	Marsupials	No growth, blastocysts with zona pellucida
Obligate diapause	Exist in every gestation, allow the birth of their offspring at favorable environment conditions; always induced by photoperiod	Prolactin withdraw for diapause; unknown ovarian factors	Mustelid carnivores	Some growth, blastocysts with zona pellucida
			Roe deer	Some growth, blastocysts without zona pellucida
			Fruit bats	Implanted but undifferentiated during delayed condition

In contrast, prolactin plays an inhibitory role for embryo implantation in macropod marsupials, such as tammar wallaby (Renfree and Shaw, 2000). Prolactin suppresses the growth and steroid secretion of corpus luteum. Prolactin is released in response to seasonal photoperiod changes or sucking stimulus. Reactivation of implantation requires a fall of prolactin secretion levels. The sucking stimulus must be withdraw for at least 3 days to allow the corpus luteum to escape from inhibition (Gordon, et al., 1987). A loss of at least three consecutive dawn pulses of prolactin secretion by experimentally manipulating photoperiod results in nonreversible reactivation of corpus luteum, thereby terminating diapause (Hinds, 1994, McConnell, et al., 1986, Renfree and Shaw, 2000). In addition, hypophysectomy, or denervation of the sucked mammary gland can remove the inhibitory effects of prolactin, resulting in implantation reactivation (Renfree and Shaw, 2000).

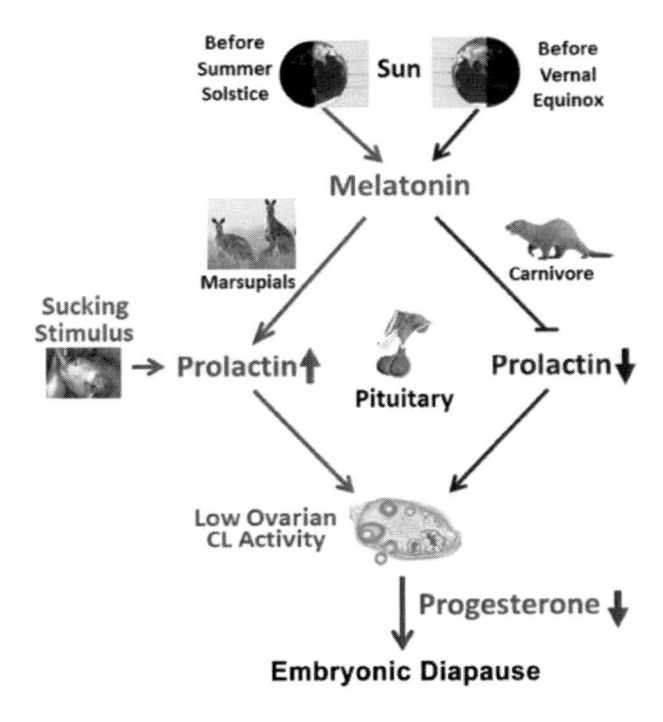

Figure 1. Seasonal and lactational control of embryonic diapause in some marsupials (tammar) and carnivores (mink).
In the tammar, diapause is controlled by the lactation from January to May. Photoperiod regulates the diapause from May to the summer solstice. The lactation and photoperiod regulation of diapause overlaps for some months. Under the photoperiod regulation before the summer solstice, prolactin is up-regulated mediated by increased melatonin secretion. Prolactin then inhibits ovarian CL activity, resulting in low progesterone output and initiation and maintenance of embryonic diapause. Sucking stimulus can also increase prolactin secretion in tammar facilitating embryonic diapause. In the mink, embryonic diapause lasts until the vernal equinox. Melatonin inhibits prolactin secretion, leading to low ovarian CL activity and low progesterone outcome, thus embryos undergo diapause. High level of prolactin is essential for the termination of diapause in the mink. CL, corpus luteum.

Under the influence of pituitary hormones, ovarian hormone steroids play a key role in the induction or termination of embryonic diapause. Ovarian corpus luteum, a factory for progesterone production, has been reported to show various biological effects in different species (Douglas, et al., 1998, Lopes, et al., 2004, Mead, 1993, Murphy, et al., 1993b, Renfree and Shaw, 2000). In mustelids, reduction of ovarian corpus luteum body size correlated with low levels of progesterone occurs during embryonic diapause (Figure1).

Significantly increase of corpus luteum volume with much higher progesterone output will be induced in response to prolactin secretion during the reinitiation of implantation. In seals, progesterone remains low during diapause and increases just before or at reactivation of the blastocyst. There is a prominent surge of estrogen just before reactivation in both phocid and otariid seals (Renfree and Shaw, 2000), although injections of either progesterone or estradiol or both fail to induce reactivation (Renfree and Shaw, 2000). In wallaby, injection of progesterone can terminate diapause, indicating a critical role of progesterone in initiation of implantation in this species (Renfree and Tyndale-Biscoe, 1973). In mice and rats, the absence of estrogen with persistent treatment of progesterone helps to maintain delayed implantation (Yoshinaga and Adams, 1966), while an injection of estrogen can terminate blastocyst dormancy. In mink and spotted skunk, plasma progesterone concentration increases during the reactivation of implantation (Stoufflet, et al., 1989). However, exogenous progesterone maintains blastocyst viability, but fails to induce reactivation or implantation. Similarly, various combinations of steroids administration fails to terminate diapause in other mustelid carnivores (Chavez and Blerkom, 1979, Foresman and Mead, 1978).

3. Cellular Aspects of Delayed and Reactivated Blastocysts

During delayed implantation, embryo suspends its development at the blastocyst stage to a dormancy state with or without zona pellucida. In macropodids (kangaroos and wallabies), embryo develops to 100-cell blastocyst with no further growth and then enters dormancy (Renfree and Shaw, 2000). It is likely that the diapause blastocyst cells are arrested in the G0 phase of the cell cycle. In honey possum, unlike macropodid marsupials, blastocyst in diapause continues slow growth then to a state when blastocyst does not expand much further (Kemp, et al., 2005). In western spotted skunk and the badger, the embryo diameter and total cell number also increase during diapause, although this proliferation is restrict to trophoblast cells (Mead, 1993, Renfree and Shaw, 2000). Resumption of blastocyst cellular activities after the termination of embryonic diapause has been found in many species. For example, in skunk, blastocyst activation from diapause coincides with increasing RNA and protein synthesis in the trophectoderm and inner cell mass (Mead, 1993, Mead and Rourke, 1985). In mice, embryonic diapause can be induced experimentally via ovariectomy on day 4 of pregnancy before preimplantation estrogen secretion and then with daily injection of progesterone from day 5. A single injection of estrogen can rapidly reinitiate the implantation process in delayed implanting mice (Yoshinaga and Adams, 1966). Employing this physiological relevant delayed implantation model, previous studies have demonstrated that dormant blastocysts show low cellular activities, such as cell proliferation, translation and metabolism (Copp, 1982, Given, 1988, Holmes and Dickson, 1975, Weitlauf, et al., 1979), while these activities can soon be elevated in blastocysts by estrogen injection in vivo (Copp, 1982, Given, 1988, Holmes and Dickson, 1975, Weitlauf, et al., 1979), coincided with the fine structure changes.

3.1. DNA Replication, RNA and Protein Synthesis

During the transition of day 4 normal blastocysts into the dormancy status in mice, blastomeres keep cell division exhibiting an increase of cell number during the first 3-4 days, and then the rate of cell division slows down with no further significant increase of cell number after day 7 of delayed implantation (Copp, 1982, Paria, et al., 1993a, Weitlauf, et al., 1979). DNA synthesis and cell proliferation can be resumed soon after the administration of estrogen. Increased cell DNA synthesis is first noted in inner cell mass at about 6 hours after estrogen administration, then to the polar and mural trophoblast cells at about 12 to 18 hours in vivo (Given and Weitlauf, 1981, Given and Weitlauf, 1982). However, apparent total cell number increase can be observed until 24 hours after estrogen injection in vivo (Copp, 1982, Weitlauf, et al., 1979). Resumption of DNA synthesis can also be observed when dormant blastocysts are cultured in basal Eagle's medium in vitro (Given and Weitlauf, 1982). However, the blastocyst under such simply culture conditions fails to acquire its physiological implantation competency (Paria, et al., 1993b), suggesting that a resumption of cell proliferation is not enough for physiological activation of dormant blastocysts.

RNA synthesis is persistent in blastocysts during delayed implantation (Chavez and Blerkom, 1979). Endogenous RNA polymerase activity can be detected for at least 5 days during delayed peroid (Chavez and Blerkom, 1979). However, nucleolus, a manufactory for ribosome RNA production, becomes more and more dense during delayed implantation (Van Blerkom, et al., 1978). Under normal pregnancy condition, a significant increase of ^{3}H-uridine incorporation reflecting RNA synthesis occurs in blastocysts between days 4 and 5 (Van Blerkom, et al., 1978). However, estrogen-induced activation of embryos in vivo does not exhibit a significant change in level of ^{3}H-uridine uptake or incorporation during the first 24 hours of activation (Chavez and Blerkom, 1979, Van Blerkom, et al., 1978).

Protein synthetic acitvity in blastocysts during delayed implantation is markly reduced compared to normal blastocysts (Van Blerkom, et al., 1978, Weitlauf, 1973a, Weitlauf, 1973b, Weitlauf and Greenwald, 1968). The protein content in the implanting blastocyst increases about 50% from 0600 am to 1600 pm on day 5 of pregnancy in mice (Weitlauf, 1973a). However, ptotein content of dormant blastocyst is not significantly different from normal day 5 0600 am blastocyst during the long-term delayed condition (Weitlauf, 1973a). The dense nucleolus in the dormant blastocysts also indicate low activatity of protein synthesis in these diapauing embryos (Van Blerkom, et al., 1978). Protein synthesis rate, by measuring uptake and incorporation rate of amino acids, increases progressively during the fisrt 24 hours after estrogen-induced reactivation (Van Blerkom, et al., 1978). However, this progressive increase of amino acid uptake and incorporation is not directly regulated by progesterone or estrogen when tested in culture (Weitlauf, 1973b). In additon, removing leucine, arginine and glucose from the culture medium can effectively suspend blastocyst outgrowth in vitro (Naeslund, 1979), while supplementation of these to culture medium can reinitiate the outgrowth (Gonzalez, et al., 2012, Naeslund, 1979), highlighting that some metabolites participate in the regulation of embryonic diapause and reactivation.

3.2. Energetic Regulation of Delayed Implantation and Reactivation

The global energetic activity of dormant blastocysts is much lower than reactivated blastocysts (Menke and McLaren, 1970). Pyruvate uptake increases within 4 hours after estrogen induced reactivation of blastocysts in vivo. In contrast, glucose uptake remains basal until 16 hours, but increases thereafter, suggesting that pyruvate is an important source of energy during early stage of blastocyst activation (Spindler, et al., 1996). By exposure of blastocysts to ^{14}C-glucose in vitro, the rate of CO_2 production by dormant blastocysts is reported to be low compared to that of implanting blastocysts at the normal time, and progressively increases 12 hours after estrogen administration (Menke and McLaren, 1970, Torbit). However, through ^{14}C-pyruvate exposure in vitro, the rate of $^{14}CO_2$ production is apparently equal in dormant and activated blastocysts (Nieder and Weitlauf, 1984). Considering the much higher pyruvate uptake in activated blastocysts (Spindler, et al., 1996), it is conceivable that the pyruvate may be utilized through other pathways, such as lactate production. Indeed, the significantly higher lactate production in activated blastocysts support this view (Nieder and Weitlauf, 1984), indicating an aerobic glycolysis during the blastocyst activation. A significantly higher lactate production has also been detected in human embryos at the blastocyst stage, compared to other stage embryos (Gott, et al., 1990). Recent evidences show that mouse blastocyst inner cell mass cells exhibit obvious Warburg effect (Hewitson and Leese, 1993, Houghton, 2006). In humans, high aerobic glycolysis activity mediated by hypoxia-inducible factor 1α is important for the transition of embryonic stem cells (ESC) to epiblast stem cells (EpiSC/hESC) (Zhou, et al., 2012), indicating a functional role of aerobic glycolysis during early stem-cell development in blastocysts. However, the role of aerobic glycolysis in trophoblast is still unknown. Increased activity of phosphofructokinase and pyruvate kinase with decreased total lactate dehydrogenase activity in dormant blastocyst suggest that glucose metabolism and glycolysis remains active during delayed implantation (Sakhuja, et al., 1982), contributing to a low CO_2 output in dormant blastocysts (Menke and McLaren, 1970). The rate of O_2 consumption by delayed blastocysts rapidly increases to nearly two fold 4 hours after estrogen injection and remains at this level 18 hours after activation for implantation (Nilsson, et al., 1982). The activity of malate dehydrogenase, an enzyme in tricarboxylic acid cycle, doesn't change significantly during first 22 hours of blastocyst activation (Sakhuja, et al., 1982). However, mitochondria cytochrome oxidase activity of delayed blastocysts exhibits no positive detection both in the trophectoderm and inner cell mass. Eight hours after estrogen injection, partial mitochondria of blastocyst cells show positive cytochrome oxidase activity, while most mitochondria show positive detection 18 hours after estrogen injection (Nilsson, et al., 1982). These findings suggest cytoplasmic glycolysis is important for both blastocyst dormancy and reactivation, although the mitochondria oxidative phosphorylation function in delayed implantation and activation remains elusive. Considering the diversity of blastocyst cell types, a detailed characterization of metabolism pathways in different cells is warranted for a better understanding of metabolic mechanisms of blastocyst dormancy and activation.

3.3. Morphological Fine Structures

Dormant and activated blastocysts also show distinct ultrastructural features from transmission and scanning electron microscope views. Trophoblast cells of dormant blastocysts are rough on the cell surface due to multiple imprints from appositional uterine epithelium cells (Holmes and Gordashko, 1980), suggesting that dormant blastocysts are held tightly within the uterine lumen. Trophoblast cells of blastocysts are expanded and bulging with distinct and raised cell borders after estrogen reactivation, when the imprints of uterine epithelium cells are no longer present on the trophoblast cell surface (Holmes and Gordashko, 1980). The expansion and swelling of trophoblast cells are found to precede the polar trophoblast cell proliferation (Bergstrom, 1972). The diapause blastocysts exhibit accumulation of lipid droplets, microfilaments, a basal lamina coating trophectodermal cells, as well as disassembly of polysomes into ribosomes. In contrast, consumption of lipid droplets, reassembly of polysomes and the accumulation of glycogen granules, large quantities of an amorphous material within the cisternae of the rough-surfaced endoplasmic reticulum are commonly seen in reactivated blastocysts (Potts, 1968, Tachi, et al., 1970, Van Blerkom, et al., 1978). Previous studies in rats also show an increase of ribosome and granular endoplasmic reticulum number in the reactivated trophoblast and inner cell mass (Wu and Meyer, 1974). The Golgi complex is large and well-developed in activated blastocysts with continuous accumulation of glycogen and the occurrence of various inclusion bodies in the trophoblast cells, but not in the cells of the inner cell mass (Wu and Meyer, 1974). These physiological changes are correlated to energy metabolism and protein synthesis during blastocyst activation. Nuclear chromatins become significantly dense with much heterochromatin at day 3 of embryonic diapause, while these heterochromatin structures rapidly disappears within just 12 hours after estrogen mediated reactivation (Van Blerkom, et al., 1978), indicating a global activation of the nuclear activities during blastocyst activation. In addition, many autophagy vacuoles with particular lysed contents are found in the trophoblast cells of day 14 dormant blastocyst, which are proved to be important for the extended longevity of dormant blastocysts in uteri during delayed implantation (Lee, et al., 2011). However, the molecular basis responsible for these characteristic physiological changes remains largely unknown.

4. MOLECULAR ASPECTS OF DELAYED AND REACTIVATED BLASTOCYSTS

The concept of uterine controlling of blastocyst dormancy and reactivation has been established by cross-species embryo transfer experiments, showing that diapause blastocysts showed implantation after transferred into the uterus of another non-diapause species, while embryonic diapause of blastocysts from non-diapausing mammals can be induced following their transfer into the delayed implanting uteri of diapause species (Chang, 1968, Ptak, et al., 2012). Therefore, uterine derived signaling molecules are crucial for induction of blastocyst diapause and reactivation for implantation. In this respect, previous studies have demonstrated that cyclooxygenase (COX) -1 and COX-2, the limiting enzyme for prostaglandin (PG) biosynthesis are not detected in the uterus during diapause in the skunk uterus, but are expressed in the trophoblast and uterine endometrium during implantation (Das, et al., 1999). Products of COX-2, PGE and PGD are implicated in upregulation of transcription of the angiogenic factors associated with implantation in the mink (Lopes, et al., 2003, Lopes, et al., 2006). Moreover, leukemia inhibitory factor (LIF), a cytokine that is essential for implantation in the mouse is expressed in the uterine glands just prior to and after implantation in the mink (Song, et al., 1998), while LIF receptors are found in uterine glands and in invading trophoblasts during implantation in the spotted skunk (Passavant, et al., 2000). Moreover, epidermal growth factor (EGF) and its receptor are present in the spotted skunk uterus both during the delay and postimplantation phases of gestation, and EGF binding activity is significantly elevated during the implantation process (Paria, et al., 1994). These findings suggest that LIF and EGF related growth factors may play a critical role during blastocyst diapause and reactivation for implantation in mustelid carnivores, such as the mink and spotted skunk. In fact, LIF or EGF can replace the nidatory estrogen pulse during delayed implantation in mice and rats, respectively (Johnson and Chatterjee, 1993, Sherwin, et al., 2004). Despite abovementioned molecular regulators involved in delayed implantation in some wild species, previous studies employing experimentally induced delayed implantation mouse model have identified several key signaling molecules essential for blastocyst dormancy and reaction, including, 4-hydroxy-E_2 (4-OH-E_2), endocannabinoids, Wnts and heparin-binding EGF-like growth factor (HB-EGF).

4.1. Estrogenic Metabolite in Blastocyst Activation during Implantation

In mice, ovarian estrogen via interacting with the nuclear estrogen receptors is essential for the establishment of uterine receptivity for implantation. However, estrogen fails to activate dormant blastocysts, although estrogen receptor is present in preimplantation mouse embryos (Hou, et al., 1996). Instead, 4-OH-E$_2$, an estrogen derived metabolite produced locally in the uterus is essential for blastocyst activation. Dormant blastocysts incubated with 4-OH-E$_2$ in culture, but not with native estrogen, show normal implantation after their transfer into the receptive uterus. Along with this finding, the receptive uterus is capable of transforming the native estrogen into catecholestrogen prior to embryo implantation (Paria, et al., 1990, Paria, et al., 1998). CYP1B1, an enzyme involved in NADPH- dependent 4-hydroxylation of 17 β-estradiol, is expressed in the uterus during the periimplantation period in mice (Paria, et al., 1998, Reese, et al., 2001). Observations of the presence of estrogen-4-hydroxylase activity in the progesterone-treated uterus and the expression of CYP1B1 mRNA in the receptive uterus strongly suggest that 4-OH-E$_2$ locally produced in the uterus directs blastocyst activation for implantation.

Further studies demonstrate that 4-OH-E$_2$, but not native estrogen, can elevate the expression of COX-2 in dormant blastocysts when cultured in vitro, suggesting that catecholestrogen initiates blastocyst activation via interacting with PG singling. In this respect, PGE$_2$ can effectively render blastocyst activation (Paria, et al., 1998), whereas COX-2 inhibitor can block the 4-OH-E$_2$-trigered blastocyst activation. In addition, dormant blastocysts cultured with cAMP can implant normally after their transfer into the receptive uterus, while inhibition of cAMP production or protein kinase A activity blocks 4-OH-E$_2$' effect (Paria, et al., 1998), suggesting that 4-OH-E$_2$ mediated activation of dormant blastocyst is cAMP-PKA dependent. It is conceivable that 4-OH-E$_2$ initiates blastocyst activation via increasing production of PGs, which in turn functions through activation of cAMP-dependent PKA pathway. However, future studies are warranted to further reveal the molecular mechanisms by which catecholestrogens exert its bioactivity and whether there is a specific receptor solely for catecholestrogens in the blastocyst and uterus during implantation. It is also equally important to address whether and how ovarian steroid hormones and their metabolites interact with other local factors to initiate embryo implantation during early pregnancy.

4.2. Endocannabinoid Signaling: A Gatekeeper for Blastocyst Dormancy and Reactivation

N-arachidonoylethanolamine (AEA, also known as anandamide) and 2-arachidonoyl glycerol (2-AG) are two major endocannabinoids, primarily functioning through two G Protein-coupled receptors, the brain-type (CB1) and spleen-type (CB2) receptors (Wang, et al., 2006). Previous studies have demonstrated that endocannabinoid signaling is an essential player in multiple early pregnancy events, including preimplantation development of embryos and their timely transport from the oviduct into the uterus, attainment of uterine receptivity, and embryo-uterine cross-talk during implantation (Das, et al., 1995, Paria, et al., 1995, Paria, et al., 2001b, Wang, et al., 2004a, Wang, et al., 2003).

Both CB1 and CB2 are expressed in preimplantation embryos in mice (Paria, et al., 1995, Yang, et al., 1996). CB1 mRNA is detected from the 4-cell to the blastocyst stages, especially in the trophectoderm of blastocyst; whereas CB2 is present from the 1-cell to the blastocyst stage, and localized in the inner cell mass. Moreover, a significantly downregulated expression of CB1 receptors has been observed in reactivated blastocysts during delayed implantation (Paria, et al., 2001b, Wang, et al., 2003), coincided with a decreased level of endogenous anandamide in the receptive uterus (Paria, et al., 1996), suggesting that a precisely regulated endocannabinoid signaling is essential for blastocyst activation and implantation. Indeed, anandamide at low concentration through CB1 receptors can confer blastocysts with implantation competency via activating extracellular signal-regulated kinase (ERK) pathway, whereas higher levels of anandamide inhibits Ca^{2+} channel activity and is detrimental to blastocyst function and reactivation (Wang, et al., 2003). Therefore, it is conceivable that critical levels of endocannabinoids produced in the uterus via interacting with appropriately expressed CB1 receptors in blastocysts synchronize blastocyst activation with uterine receptivity for implantation. Aberrant levels of uterine endocannabinoids and/or blastocyst CB1 receptors interfere with the normal embryo-uterine interactions during implantation, resulting in early pregnancy loss. It is worthy to note that spontaneous pregnancy losses are associated with elevated anandamide levels in women (Habayeb, et al., 2008, Maccarrone, et al., 2002, Maccarrone, et al., 2000), reinforcing that endocannabinoid signaling is a gatekeeper for blastocyst dormancy and reactivation during implantation.

4.3. Activation of Wnt β-catenin Signaling Directs Blastocyst Differentiation Conducive for Implantation

Since embryo implantation involves an intricate interplay between the blastocyst and uterus, Wnt signaling as a highly conserved mediator of reciprocal cell-cell communications during organogenesis (van Amerongen and Nusse, 2009), is likely to participate in the cross-talk during implantation (Chen, et al., 2009). In fact, previous studies have revealed unique expression profiles of multiple Wnt genes and their pathway membranes in early embryos and uteri during the peri-implantation period in mice (Hamatani, et al., 2004a, Kemp, et al., 2005, Mohamed, et al., 2005, Paria, et al., 2001a, Wang, et al., 2004b), suggesting that Wnt signaling is operative during early pregnancy.

Although various studies have demonstrated that β-catenin-dependent canonical Wnt signaling is dispensable for blastocyst formation, it is required for normal blastocyst function during implantation. Upon crossbred with wild-type males, female mice with conditional deletion of β-catenin in oocytes produce reduced number of pups in comparison with those of wild-type (De Vries, et al., 2004), suggesting an impaired implantation for those lacking maternal β-catenin. Using the strategy of DKK1-mediated functional inhibition of nuclear β-catenin signaling and small molecule inhibitors of Wnt signaling, previous study has demonstrated that silencing canonical Wnt/β-catenin signaling does not adversely affect the uterine preparation for receptivity, but remarkably blocks blastocyst competency for implantation in mice (Xie, et al., 2008). The significance of this pathway in blastocysts is further evidenced from the findings in delayed implantation models, showing that the activity of nuclear β-catenin signaling distinguishes blastocyst dormancy from activation. Moreover, Wnt3a is able to induce intracellular accumulation and nuclear translocation of β-catenin in trophectoderm cells and concurrently induces the expression of peroxisome proliferator-activated receptor δ, a nuclear receptor for prostacyclin. Pharmological gain of function study further reveals that canonical Wnt signaling synergizes with PG signaling to confer blastocyst competency for implantation. This is consistent with early findings that expression of COX-2 is substantially upregulated in activated blastocysts after treatment with catecholestrogen (Paria, et al., 1998). Nonetheless, these findings constitute direct evidence that Wnt signaling is at least one pathway determining blastocyst competency for implantation.

4.4. A Global Molecular Signature Distinguishes Blastocyst Dormancy versus Activation

To better understand molecular mechanisms governing embryonic diapause and reactivation, previous study has employed microarray technique to analyze the global gene expression profile of mouse dormant versus activated blastocysts using delayed implantation model (Hamatani, et al., 2004b). Total 229 differentially expressed genes among about 20,000 genes have been identified between dormant and activated blastocysts. The main functional categories of altered genes include cell-cycle, calcium signaling, chromatin remodeling and energy-metabolic pathways, and so on. Resumptions of cell proliferation in reactivated blastocysts are identified, consistent with previous observations of increased DNA replication in reactivated blastocysts (Given and Weitlauf, 1981, Given and Weitlauf, 1982). Cell-cycle regulatory molecules and genes involved in DNA replication, such as Ccnd1, Ccne1, Ccne2, Cdc6, Cdc45, Orc1 and Mcm5, are found to be up-regulated in activated blastocysts. Moreover, Brca1, a known estrogen-response gene, is up-regulated in activated blastocysts, coordinated with down-regulation of p21. With respect to energy metabolism, expression levels of phosphofructokinase 1, phosphoglycerate kinase 1, enolase 1, pyruvate kinase 2 and lactate dehydrogenase 1 genes are significantly increased in activated blastocysts, indicating a strong glycolytic activity in blastocysts upon reactivation. Moreover, genes involved in calcium signaling important for preimplantation embryo development and blastocyst adhesion (Stachecki and Armant, 1996, Wang, et al., 2003, Wang, et al., 2002) are differentially expressed in dormant versus activated blastocysts. For example, expressions of Pik3c2a, Pitpn, Vav3, Pr1 and Etohd4/Fgd6 are up-regulated in activated blastocysts, while Tmr4, Pepp1, S100a13, Tpd52l1 and Fer1l3/myoferlin are down-regulated during blastocyst activation (Hamatani, et al., 2004b). Despite all these findings, it remains to explore how differential signaling molecules direct the process of blastocyst dormancy and reactivation.

4.5. HB-EGF: An Early Messenger of Blastocyst-uterine Dialogue during Implantation

Since the nature of embryo implantation is an intricate physical and physiological interaction between the embryo and uterus, it is conceivable that embryo-derived signaling molecules upon activation are likely to be involved in this two-way dialogue during implantation. In this regard, HB-EGF has been demonstrated to be so far the earliest signaling molecule involved in embryo-uterine cross-talk during the periimplantation period (Lim and Dey, 2009). HB-EGF is expressed in the luminal epithelium at the site of blastocyst apposition approximate 6 hours prior to blastocyst attachment reaction in mice (Das, et al., 1994). This spatiotemporal HB-EGF expression is dependent on the presence of blastocysts, since HB-EGF is not expressed in the pseudopregnant mouse uterus or at other epithelial cells outside of the blastocyst apposition site on day 4 afternoon (Das, et al., 1994, Xie, et al., 2007). Moreover, an increasing expression of ErbB1 and ErbB4 receptors and their ligand binding activity have been observed in reactivated blastocysts in delayed implantation model (Das, et al., 1997, Paria, et al., 1993b, Raab, et al., 1996). There is also evidence that HB-EGF can facilitate blastocyst hatching and trophoblast cell differentiation and outgrowth via activating intracellular Ca^{2+} signaling and protein kinase C pathway (Wang, et al., 2000).

Moreover, a significant increase of HB-EGF expression has been observed in the trophectoderm cell of activated blastocysts (Hamatani, et al., 2004b). Most interestingly, blastocyst-size Affi-gel blue beads presoaked with HB-EGF protein can induce its own gene expression in uterine cells surrounding the beads after intraluminal transfer and render a blue reaction, similar as that induced by normal blastocysts (Hamatani, et al., 2004b, Paria, et al., 2001a). This observation suggests that an auto-induction loop of HB-EGF between the implanting blastocyst and uteri acting in a paracrine and juxtacrine manner mediates the two-way dialogue during the initiation of implantation. Deficiency of either ErbB1 or ErbB4 in embryos doesn't affect early implantation events (Gassmann, et al., 1995, Threadgill, et al., 1995), indicating potential compensatory effects of ErbB receptors or heparin sulfate proteoglycan molecules during implantation. However, HB-EGF null mutant mice exhibit a deferral of implantation window accompanied with compromised term pregnancy (Xie, et al., 2007), highlighting the necessity of HB-EGF signaling for normal embryo implantation. In humans, HB-EGF is expressed at high levels in the receptive endometrium (Birdsall, et al., 1996, Leach, et al., 1999, Stavreus-Evers, et al., 2002, Yoo, et al., 1997) and its receptor ErbB4 is localized on the surface of the trophectoderm in peri-implantation human blastocysts (Chobotova, et al., 2002), indicating that HB-EGF signaling also mediates the trophectoderm-uterine epithelium interaction during implantation in humans.

5. How to Select a Good Embryo for Transfer in Human Clinical Practice

With the advance of our knowledge on the blastocyst function and implantation, it has been hoped that all new discoveries from basic researches using animal models can be quickly translated into the clinical practice. However, owing to the ethics restriction and other issues, this translational progress remains slow. A gradually increased population facing fertility problem becomes a world-wide challenge for reproductive biologists and physicians. In this respect, application of in vitro fertilization and embryo transfer approach (IVF-ET) has helped an increasing number of infertile couples. Significant improvement of pregnancy success rate has been achieved by blastocyst transfer approach in IVF (even to 60%) compared to cleavage embryo transfer (about 10-15%) (Gardner, 1998, Gardner and Lane, 1997), highlighting that quality of embryos determines the fate of implantation and thus term pregnancy. Therefore, many strategies have been developed and applied for the evaluation human blastocyst viability (Table 2).

Table 2. Non-invasive and invasive assessments of human embryo viability

Non-invasive assessments	Invasive assessments
Gross morphology: Zona pellucida Embryo developmental rate and status Cell number of trophectoderm and inner cell mass	**Chromosome status:** Aneuploidy detected by FISH or CGH **Nuclear status:** Mitotic, pycnotic or apoptotic index analyzed by confocal microscope
Embryo secretome: Amino acids: utilization and turnover Specific secreted factors: PAF, a 85 kDa protein Metabolic profile: Metabolites related to oxidative stress analyzed by NIR and Raman spectroscopy	**Mitochondria and lysosome features:** Immunostaining or electron microscope **Cytoskeleton features:** Fluorescence staining

5.1. Non-invasive Evaluation of Human Blastocysts for Transfer

Non-invasive assessment of blastocyst quality by the morphological grade has improved the implantation rate in patients undergoing IVF treatment (Gardner, et al., 2000). A morphologically grading system of human blastocysts takes account of the thickness of zona pellucida, the embryo development rate and the morphological qualities of inner cell mass and trophectoderm (Gardner, et al., 2000). Top-score blastocysts show 50% implantation rate after single embryo transfer to women about 32-33 year old (Gardner, et al., 2000). Quick and convenient, noninvasive morphological assessment of blastocyst quality is currently most widely used (Van Soom, et al., 2001). However, the multi-pregnant rate in the group of women transferred with two top-score blastocysts achieved at 61%, which calls for more accurate assessments of blastocyst quality to achieve the goal of single blastocyst transfer.

Non-invasive investigation of embryo metabolites has been previously attempted to assess embryo viability (Conaghan, et al., 1993b, Gardner, et al., 2001, Leese, et al., 2008, Turner, et al., 1994, Urbanski, et al., 2008). Pyruvate, one of the major energy sources reported to be highly utilized during preimplantation embryo development (Conaghan, et al., 1993a, Gott, et al., 1990), has been shown to be utilized inefficiently in embryos that implant successfully (Turner, et al., 1994). The wide variation of pyruvate uptake suggests that pyruvate maybe not a good marker for embryo quality. Glucose uptake is increased significantly at the transition from morula to blastocyst, and greater glucose uptake is also observed in human blastocysts ranking higher morphological grade (Gardner, et al., 2001). However, there is also evidence showing no significant relativity between glucose uptake and embryo viability (Botros, et al., 2008, Jones, et al., 2001). Thus, similar to pyruvate, it is also delusive whether glucose consumption can predict embryo viability. Moreover, assessment of amino acid taken over or secretion has been applied to evaluate embryo developmental viability (Brison, et al., 2004, Houghton, et al., 2002). The profile of Ala, Arg, Gln, Met and Asn flux can predict blastocyst potentiality at >95% (Houghton, et al., 2002). In addition, the turnover of Asn, Gly and Leu is significantly correlated to clinical pregnancy and live birth, independent of known predictors, such as female age, basal levels of follicle stimulating hormones (FSH), embryo cell number and embryo morphological grade (Brison, et al., 2004). Embryos with greater viability tend to have a lower or quieter amino acid metabolism than those that arrest (Brison, et al., 2004, Sturmey, et al., 2009), supporting the "quiet embryo hypothesis" that non-viable embryos are more metabolically active than developmentally competent cleavage embryos (Leese, 2002).

Embryo metabolomics via Near-infrared (NIR) Raman Spectroscopy analysis has been recently developed as a viability index system for embryo implantation capacity (Seli, et al., 2007). This metabolic profile approach for embryo culture medium doesn't identify or quantitate specific metabolites, but a group of metabolites related to oxidative stress such as -CH, -OH, -NH groups. The mean viability score of embryos that are implanted and further given to birth is much higher in embryos than those fails to implant (Dominguez, et al., 2008, Hardarson, et al., 2008, Scott, et al., 2008, Seli, et al., 2007). These findings reveal an association between oxidative stress and embryo implantation competency. Moreover, the metabolic profile of embryos is independent of embryo morphology (Hardarson, et al., 2008, Scott, et al., 2008, Seli, et al., 2007), indicating that a combined consideration of metabolic viability index with embryo morphology grade may improve successful rate of single embryo transfer.

In addition to embryo metabolomics, embryo secretome has also been used for non-invasive evaluation of blastocyst implantation ability. By analyzing the secretome of human implanted or non-implanted blastocysts, Chemokine (C-X-C motif) ligand 13 and granulocyte-macrophage colony stimulating factor have been identified significantly decreased in culture medium of implanted blastocysts compared to non-implanted blastocysts (Dominguez, et al., 2008). Moreover, detection of soluble leukocyte antigen-G (HLA-G) in day 3 human IVF embryo culture medium tends to be associated with higher pregnant rate, although a portion of embryos that do not secrete HLA-G still show pregnancy (Noci, et al., 2005, Sher, et al., 2005). An 8.5 kDa protein has also been identified correlated with the developmental potential of embryos (Katz-Jaffe and Gardner, 2007, Katz-Jaffe, et al., 2006). The best candidate for this 8.5 kDa protein is ubiquitin, a component of proteasome. In addition, 1-o-alkyl-2-acetyl-sn-gylcero-3-phosphocholine (PAF), one of the earliest trophins produced and secreted by embryos of all mammalian species, is further verified in human embryos secretome (O'Neill, 2005).

5.2. Invasive Assessments of Human Blastocyst Viability

Invasive assessment of embryos involves investigation of chromosomal status, the nuclear status, the mitochondrial status, the cytoskeleton, and the cell organelles (Van Soom, et al., 2001). Invasive assessments can provide genetic and cellular disorders directly. Aneuploidy is a kind of chromosomal anomaly frequently happened in in-vitro culture embryos. In humans, chromosomal anomalies frequency increases with advancing of female age (Fishel, et al., 2010, Fragouli, et al., 2011, Fragouli, et al., 2006). Fluorescent in situ hybridization (FISH), comparative genomic hybridization (CGH) and microarray-CGH (aCGH) methods are used in diagnosing aneuploidy (Fragouli, et al., 2011). The mitochondria and cytoskeleton can also be assessed by fluorescent staining (Acton, et al., 2004, Wakefield, et al., 2011, Warner, et al., 2004). Some other parameters are also used for assess the manipulation of embryos. Epigenetic DNA methylation analysis reveals that intracytoplasmic sperm injection (ICSI) does not increase incidence of epigenetic errors in human embryos (Santos, et al., 2010). Biopsy with a very small group of trophectoderm cells from blastocysts provides possibilities of many manipulations, such as PCR, immunostaining and preimplantation genetic diagnose (PGD). However, it must be evaluated for the embryo viability, pregnancy potential and offspring long-term health. In mice, embryo biopsy at or before 8-cell stage can affect fetal viability (Krzyminska, et al., 1990), coupled with potential high risks of neurodegenerative disorders in the offspring (Yu, et al., 2009). These risks should be considered when biopsy is used in human IVF.

Collectively, morphological assessment of blastocysts is still the most convenient and widely-used approach to date. Evaluation of blastocyst quality according to morphological characters is subjective based on different manipulators, thus results in large variations of implantation rate in IVF. Although implantation rate have been improved by various non-invasive and invasive assessments of human embryos, the mechanisms determining embryo quality need to be understood in depth to develop reliable indicators for human blastocyst quality.

CONCLUSION

Blastocyst activation, a process for the blastocyst to achieve implantation competency is equally important as attainment of uterine receptivity for the success of embryo implantation. While a wide range of regulatory molecules have been identified as essential players in conferring uterine receptivity in both laboratory animal models and humans (Dey, et al., 2004, Lim and Wang, 2010, Tranguch, et al., 2005, Wang and Dey, 2006), it remains largely unknown how blastocysts achieve implantation competency. During the past decade, increasing attentions have been paid to search for biomarkers predicting human blastocyst implantation rate in IVF. However, the progress of these efforts remains slow owing to methodological and material limitation and ethic restriction in humans. With the advance of microscale proteomics and genomics (Aghajanova, et al., 2012, Dominguez, et al., 2010, Katz-Jaffe, et al., 2006, Tang, et al., 2009), it becomes possible to characterize embryonic contribution during implantation. Moreover, trophectoderm cells are traditionally the central cell-types in various blastocyst functional assays, while the potential contribution of inner cell mass to the blastocyst function and activation for implantation has been chronically neglected. In fact, a significant activation of inner cell mass has been observed in blastocysts a few hours after reactivation from diapause (Van Blerkom, et al., 1978). Moreover, in human IVF, cell quality and shape of inner cell mass will greatly affect the implantation rate (Richter, et al., 2001). It is believed that a coordinated interaction between the trophectoderm and inner cell mass is essential for normal blastocyst function during implantation. Therefore, within the implantation niche, blastocyst trophectoderm cells perform multiple cell-cell interactions with the inner cell mass as well as the uterine luminal epithelium to obtain implantation competency and initiate implantation. A better understanding on the nature of these dialogues will help to develop new strategies for IVF assessment and treatment.

Acknowledgment

Work incorporated in this article was partially supported by the National Basic Research Program of China (2011CB944401, 2010CB945002), and the National Natural Science Foundation (30825015, 81130009).

References

Acton, B. M., Jurisicova, A., Jurisica, I. and Casper, R. F. (2004). Alterations in mitochondrial membrane potential during preimplantation stages of mouse and human embryo development. *Mol Hum Reprod,* 10, 23-32.

Aghajanova, L., Shen, S., Rojas, A. M., Fisher, S. J., Irwin, J. C. and Giudice, L. C. (2012). Comparative transcriptome analysis of human trophectoderm and embryonic stem cell-derived trophoblasts reveal key participants in early implantation. *Biol Reprod*, 86, 1-21.

Bergstrom, S. (1972). Scanning electron microscopy of ovoimplantation. *Arch Gynakol*, 212, 285-307.

Birdsall, M. A., Hopkisson, J. F., Grant, K. E., Barlow, D. H. and Mardon, H. J. (1996). Expression of heparin-binding epidermal growth factor messenger RNA in the human endometrium. *Mol Hum Reprod*, 2, 31-34.

Botros, L., Sakkas, D. and Seli, E. (2008). Metabolomics and its application for non-invasive embryo assessment in IVF. *Mol Hum Reprod*, 14, 679-690.

Brison, D. R., Houghton, F. D., Falconer, D., Roberts, S. A., Hawkhead, J., Humpherson, P. G., Lieberman, B. A. and Leese, H. J. (2004). Identification of viable embryos in IVF by non-invasive measurement of amino acid turnover. *Hum Reprod*, 19, 2319-2324.

Chang, M. C. (1968). Reciprocal Insemination and Egg Transfer between Ferrets and Mink. *Journal of Experimental Zoology*, 168, 49-&.

Chavez, D. J. and Blerkom, J. V. (1979). Persistence of embryonic RNA synthesis during facultative delayed implantation in the mouse. *Dev Biol*, 70, 39-49.

Chen, Q., Zhang, Y., Lu, J., Wang, Q., Wang, S., Cao, Y., Wang, H. and Duan, E. (2009). Embryo-uterine cross-talk during implantation: the role of Wnt signaling. *Mol Hum Reprod*, 15, 215-221.

Chobotova, K., Spyropoulou, I., Carver, J., Manek, S., Heath, J. K., Gullick, W. J., Barlow, D. H., Sargent, I. L. and Mardon, H. J. (2002). Heparin-binding epidermal growth factor and its receptor ErbB4 mediate implantation of the human blastocyst. *Mech Dev*, 119, 137-144.

CK, W. (1940). The experimental shortening of delayed pregnancy in the albino rat. *Anatomical Record*, 77 31–48,

Cockburn, K. and Rossant, J. (2010). Making the blastocyst: lessons from the mouse. *J Clin Invest*, 120, 995-1003.

Conaghan, J., Handyside, A. H., Winston, R. M. and Leese, H. J. (1993a). Effects of pyruvate and glucose on the development of human preimplantation embryos in vitro. *J Reprod Fertil*, 99, 87-95.

Conaghan, J., Hardy, K., Handyside, A. H., Winston, R. M. and Leese, H. J. (1993b). Selection criteria for human embryo transfer: a comparison of pyruvate uptake and morphology. *J Assist Reprod Genet*, 10, 21-30.

Copp, A. J. (1982). Effect of implantational delay on cellular proliferation in the mouse blastocyst. *J Reprod Fertil*, 66, 681-685.

Daniel, J. F. (1910). Observations on the period of gestation in white mice. *Journal of Experimental Zoology,* 9, 865-870.

Das, S. K., Wang, X. N., Paria, B. C., Damm, D., Abraham, J. A., Klagsbrun, M., Andrews, G. K. and Dey, S. K. (1994). Heparin-binding EGF-like growth factor gene is induced in the mouse uterus temporally by the blastocyst solely at the site of its apposition: a possible ligand for interaction with blastocyst EGF-receptor in implantation. *Development,* 120, 1071-1083.

Das, S. K., Paria, B. C., Chakraborty, I. and Dey, S. K. (1995). Cannabinoid ligand-receptor signaling in the mouse uterus. *Proc Natl Acad Sci U S A,* 92, 4332-4336.

Das, S. K., Das, N., Wang, J., Lim, H., Schryver, B., Plowman, G. D. and Dey, S. K. (1997). Expression of betacellulin and epiregulin genes in the mouse uterus temporally by the blastocyst solely at the site of its apposition is coincident with the "window" of implantation. *Dev Biol,* 190, 178-190.

Das, S. K., Wang, J., Dey, S. K. and Mead, R. A. (1999). Spatiotemporal expression of cyclooxygenase 1 and cyclooxygenase 2 during delayed implantation and the periimplantation period in the Western spotted skunk. *Biol Reprod,* 60, 893-899.

De Vries, W. N., Evsikov, A. V., Haac, B. E., Fancher, K. S., Holbrook, A. E., Kemler, R., Solter, D. and Knowles, B. B. (2004). Maternal beta-catenin and E-cadherin in mouse development. *Development,* 131, 4435-4445.

Dey, S. K., Lim, H., Das, S. K., Reese, J., Paria, B. C., Daikoku, T. and Wang, H. (2004). Molecular cues to implantation. *Endocr Rev,* 25, 341-373.

Dey, S. K. (2005). Reproductive biology: fatty link to fertility. *Nature,* 435, 34-35.

Dominguez, F., Gadea, B., Esteban, F. J., Horcajadas, J. A., Pellicer, A. and Simon, C. (2008). Comparative protein-profile analysis of implanted versus non-implanted human blastocysts. *Hum Reprod,* 23, 1993-2000.

Dominguez, F., Gadea, B., Mercader, A., Esteban, F. J., Pellicer, A. and Simon, C. (2010). Embryologic outcome and secretome profile of implanted blastocysts obtained after coculture in human endometrial epithelial cells versus the sequential system. *Fertil Steril,* 93, 774-782 e771.

Douglas, D. A., Song, J. H., Houde, A., Cooke, G. M. and Murphy, B. D. (1997). Luteal and placental characteristics of carnivore gestation: expression of genes for luteotrophic receptors and steroidogenic enzymes. *J Reprod Fertil Suppl,* 51, 153-166.

Douglas, D. A., Song, J. H., Moreau, G. M. and Murphy, B. D. (1998). Differentiation of the corpus luteum of the mink (Mustela vison): mitogenic and steroidogenic potential of luteal cells from embryonic diapause and postimplantation gestation. *Biol Reprod,* 58, 1163-1169.

Fishel, S., Gordon, A., Lynch, C., Dowell, K., Ndukwe, G., Kelada, E., Thornton, S., Jenner, L., Cater, E., Brown, A. and Garcia-Bernardo, J. (2010). Live birth after polar body array comparative genomic hybridization prediction of embryo ploidy-the future of IVF? *Fertil Steril,* 93, 1006 e1007-1006 e1010.

Foresman, K. R. and Mead, R. A. (1978). Luteal control of nidation in the Ferret (Mustela putorius). *Biol Reprod,* 18, 490-496.

Fragouli, E., Wells, D., Thornhill, A., Serhal, P., Faed, M. J., Harper, J. C. and Delhanty, J. D. (2006). Comparative genomic hybridization analysis of human oocytes and polar bodies. *Hum Reprod,* 21, 2319-2328.

Fragouli, E., Alfarawati, S., Daphnis, D. D., Goodall, N. N., Mania, A., Griffiths, T., Gordon, A. and Wells, D. (2011). Cytogenetic analysis of human blastocysts with the use of FISH, CGH and aCGH: scientific data and technical evaluation. *Hum Reprod,* 26, 480-490.

Franklin, B. C. (1958). *Studies on the effects of progesterone on the physiology of reproduction in the mink,* Mustela vison.

Gales, N. J., Williamson, P., Higgins, L. V., Blackberry, M. A. and James, I. (1997). Evidence for a prolonged postimplantation period in the Australian sea lion (Neophoca cinerea). *J Reprod Fertil,* 111, 159-163.

Gardner, D. K. and Lane, M. (1997). Culture and selection of viable blastocysts: a feasible proposition for human IVF? *Hum Reprod Update,* 3, 367-382.

Gardner, D. K. (1998). Development of serum-free media for the culture and transfer of human blastocysts. *Hum Reprod,* 13 Suppl 4, 218-225.

Gardner, D. K., Lane, M., Stevens, J., Schlenker, T. and Schoolcraft, W. B. (2000). Blastocyst score affects implantation and pregnancy outcome: towards a single blastocyst transfer. *Fertil Steril,* 73, 1155-1158.

Gardner, D. K., Lane, M., Stevens, J. and Schoolcraft, W. B. (2001). Noninvasive assessment of human embryo nutrient consumption as a measure of developmental potential. *Fertil Steril,* 76, 1175-1180.

Gassmann, M., Casagranda, F., Orioli, D., Simon, H., Lai, C., Klein, R. and Lemke, G. (1995). Aberrant neural and cardiac development in mice lacking the ErbB4 neuregulin receptor. *Nature,* 378, 390-394.

Given, R. L. and Weitlauf, H. M. (1981). Resumption of DNA synthesis during activation of delayed implanting mouse blastocysts. *J Exp Zool,* 218, 253-259.

Given, R. L. and Weitlauf, H. M. (1982). Resumption of DNA synthesis in delayed implanting mouse blastocysts during activation in vitro. *J Exp Zool,* 224, 111-114.

Given, R. L. (1988). DNA synthesis in the mouse blastocyst during the beginning of delayed implantation. *J Exp Zool,* 248, 365-370.

Gonzalez, I. M., Martin, P. M., Burdsal, C., Sloan, J. L., Mager, S., Harris, T. and Sutherland, A. E. (2012). Leucine and arginine regulate trophoblast motility through mTOR-dependent and independent pathways in the preimplantation mouse embryo. *Dev Biol,* 361, 286-300.

Gordon, K., Renfree, M. B., Short, R. V. and Clarke, I. J. (1987). Hypothalamo-pituitary portal blood concentrations of beta-endorphin during suckling in the ewe. *J Reprod Fertil,* 79, 397-408.

Gott, A. L., Hardy, K., Winston, R. M. and Leese, H. J. (1990). Non-invasive measurement of pyruvate and glucose uptake and lactate production by single human preimplantation embryos. *Hum Reprod,* 5, 104-108.

Habayeb, O. M., Taylor, A. H., Finney, M., Evans, M. D. and Konje, J. C. (2008). Plasma anandamide concentration and pregnancy outcome in women with threatened miscarriage. *JAMA,* 299, 1135-1136.

Hamatani, T., Carter, M. G., Sharov, A. A. and Ko, M. S. (2004a). Dynamics of global gene expression changes during mouse preimplantation development. *Dev Cell,* 6, 117-131.

Hamatani, T., Daikoku, T., Wang, H., Matsumoto, H., Carter, M. G., Ko, M. S. and Dey, S. K. (2004b). Global gene expression analysis identifies molecular pathways distinguishing blastocyst dormancy and activation. *Proc Natl Acad Sci U S A,* 101, 10326-10331.

Hardarson, T., Tucker, M., Seli, E., Botros, L., Roos, P. and Sakkas, D. (2008). Non-invasive metabolic profiling of day 5 embryo culture media adds to the discriminatory power of blastocyst culture for single embryo transfer. *Fertil Steril,* 90, S77-S77.

Hewitson, L. C. and Leese, H. J. (1993). Energy metabolism of the trophectoderm and inner cell mass of the mouse blastocyst. *J Exp Zool,* 267, 337-343.

Hinds, L. A. (1994). Prolactin, a hormone for all seasons: endocrine regulation of seasonal breeding in the Macropodidae. *Oxford Reviews of Reproductive Biology-Cloth Edition,* 16, 249-302.

Holmes, P. V. and Dickson, A. D. (1975). Temporal and spatial aspects of oestrogen-induced RNA, protein and DNA synthesis in delayed-implantation mouse blastocysts. *J Anat,* 119, 453-459.

Holmes, P. V. and Gordashko, B. J. (1980). Evidence of prostaglandin involvement in blastocyst implantation. *J Embryol Exp Morphol,* 55, 109-122.

Hou, Q., Paria, B. C., Mui, C., Dey, S. K. and Gorski, J. (1996). Immunolocalization of estrogen receptor protein in the mouse blastocyst during normal and delayed implantation. *Proc Natl Acad Sci U S A,* 93, 2376-2381.

Houghton, F. D., Hawkhead, J. A., Humpherson, P. G., Hogg, J. E., Balen, A. H., Rutherford, A. J. and Leese, H. J. (2002). Non-invasive amino acid turnover predicts human embryo developmental capacity. *Hum Reprod,* 17, 999-1005.

Houghton, F. D. (2006). Energy metabolism of the inner cell mass and trophectoderm of the mouse blastocyst. *Differentiation,* 74, 11-18.

Johnson, D. C. and Chatterjee, S. (1993). Embryo implantation in the rat uterus induced by epidermal growth factor. *J Reprod Fertil,* 99, 557-559.

Jones, G. M., Trounson, A. O., Vella, P. J., Thouas, G. A., Lolatgis, N. and Wood, C. (2001). Glucose metabolism of human morula and blastocyst-stage embryos and its relationship to viability after transfer. *Reprod Biomed Online,* 3, 124-132.

Katz-Jaffe, M. G., Schoolcraft, W. B. and Gardner, D. K. (2006). Analysis of protein expression (secretome) by human and mouse preimplantation embryos. *Fertil Steril,* 86, 678-685.

Katz-Jaffe, M. G. and Gardner, D. K. (2007). Embryology in the era of proteomics. *Theriogenology,* 68 Suppl 1, S125-130.

Kemp, C., Willems, E., Abdo, S., Lambiv, L. and Leyns, L. (2005). Expression of all Wnt genes and their secreted antagonists during mouse blastocyst and postimplantation development. *Dev Dyn,* 233, 1064-1075.

King, H. D. (1913). Some anomalies in the gestation of the albino rat (Mus Norvegicus albinus). *Biological Bulletin,* 24, 377-391.

Krzyminska, U. B., Lutjen, J. and O'Neill, C. (1990). Assessment of the viability and pregnancy potential of mouse embryos biopsied at different preimplantation stages of development. *Hum Reprod,* 5, 203-208.

Leach, R. E., Khalifa, R., Ramirez, N. D., Das, S. K., Wang, J., Dey, S. K., Romero, R. and Armant, D. R. (1999). Multiple roles for heparin-binding epidermal growth factor-like growth factor are suggested by its cell-specific expression during the human endometrial cycle and early placentation. *J Clin Endocrinol Metab*, 84, 3355-3363.

Lee, J. E., Oh, H. A., Song, H., Jun, J. H., Roh, C. R., Xie, H., Dey, S. K. and Lim, H. J. (2011). Autophagy regulates embryonic survival during delayed implantation. *Endocrinology*, 152, 2067-2075.

Leese, H. J. (2002). Quiet please, do not disturb: a hypothesis of embryo metabolism and viability. *Bioessays*, 24, 845-849.

Leese, H. J., Baumann, C. G., Brison, D. R., McEvoy, T. G. and Sturmey, R. G. (2008). Metabolism of the viable mammalian embryo: quietness revisited. *Mol Hum Reprod*, 14, 667-672.

Lim, H. J. and Dey, S. K. (2009). HB-EGF: a unique mediator of embryo-uterine interactions during implantation. *Exp Cell Res*, 315, 619-626.

Lim, H. J. and Wang, H. (2010). Uterine disorders and pregnancy complications: insights from mouse models. *J Clin Invest*, 120, 1004-1015.

Lopes, F. L., Desmarais, J., Gevry, N. Y., Ledoux, S. and Murphy, B. D. (2003). Expression of vascular endothelial growth factor isoforms and receptors Flt-1 and KDR during the peri-implantation period in the mink, Mustela vison. *Biol Reprod*, 68, 1926-1933.

Lopes, F. L., Desmarais, J. A. and Murphy, B. D. (2004). Embryonic diapause and its regulation. *Reproduction*, 128, 669-678.

Lopes, F. L., Desmarais, J., Ledoux, S., Gevry, N. Y., Lefevre, P. and Murphy, B. D. (2006). Transcriptional regulation of uterine vascular endothelial growth factor during early gestation in a carnivore model, Mustela vison. *J Biol Chem*, 281, 24602-24611.

Ma, W. G., Song, H., Das, S. K., Paria, B. C. and Dey, S. K. (2003). Estrogen is a critical determinant that specifies the duration of the window of uterine receptivity for implantation. *Proc Natl Acad Sci U S A*, 100, 2963-2968.

Maccarrone, M., Valensise, H., Bari, M., Lazzarin, N., Romanini, C. and Finazzi-Agro, A. (2000). Relation between decreased anandamide hydrolase concentrations in human lymphocytes and miscarriage. *Lancet*, 355, 1326-1329.

Maccarrone, M., Bisogno, T., Valensise, H., Lazzarin, N., Fezza, F., Manna, C., Di Marzo, V. and Finazzi-Agro, A. (2002). Low fatty acid amide hydrolase and high anandamide levels are associated with failure to achieve an ongoing pregnancy after IVF and embryo transfer. *Mol Hum Reprod*, 8, 188-195.

Marois, G. (1982). [Inhibition of nidation in mice by modification of the environment and pheromones. Re-establishment by prolactin and thioproperazine]. *Ann Endocrinol* (Paris), 43, 41-52.

McConnell, S. J., Tyndale-Biscoe, C. H. and Hinds, L. A. (1986). Change in duration of elevated concentrations of melatonin is the major factor in photoperiod response of the tammar, Macropus eugenii. *J Reprod Fertil,* 77, 623-632.

McLaren, A. (1969). Stimulus and response during early pregnancy in the mouse. *Nature,* 221, 739-741.

Mead, R. A. and Rourke, A. W. (1985). Accumulation of RNA in blastocysts during embryonic diapause and the periimplantation period in the western spotted skunk. *J Exp Zool,* 235, 65-70.

Mead, R. A. (1993). Embryonic diapause in vertebrates. *J Exp Zool,* 266, 629-641.

Menke, T. M. and McLaren, A. (1970). Carbon dioxide production by mouse blastocysts during lactational delay of implantation or after ovariectomy. *J Endocrinol,* 47, 287-294.

Mohamed, O. A., Jonnaert, M., Labelle-Dumais, C., Kuroda, K., Clarke, H. J. and Dufort, D. (2005). Uterine Wnt/beta-catenin signaling is required for implantation. *Proc Natl Acad Sci* U S A, 102, 8579-8584.

Murphy, B. D. and James, D. A. (1974). The effects of light and sympathetic innervation to the head on nidation in mink. J Exp Zool, 187, 267-276.

Murphy, B. D., Concannon, P. W., Travis, H. F. and Hansel, W. (1981). Prolactin: the hypophyseal factor that terminates embryonic diapause in mink. *Biol Reprod,* 25, 487-491.

Murphy, B. D. (1983). Precocious induction of luteal activation and termination of delayed implantation in mink with the dopamine antagonist pimozide. *Biol Reprod,* 29, 658-662.

Murphy, B. D. and Rajkumar, K. (1985). Prolactin as a luteotrophin. *Can J Physiol Pharmacol,* 63, 257-264.

Murphy, B. D., DiGregorio, G. B., Douglas, D. A. and Gonzalez-Reyna, A. (1990). Interactions between melatonin and prolactin during gestation in mink (Mustela vison). *J Reprod Fertil,* 89, 423-429.

Murphy, B. D., Rajkumar, K., Gonzalez Reyna, A. and Silversides, D. W. (1993a). Control of luteal function in the mink (Mustela vison). *J Reprod Fertil Suppl,* 47, 181-188.

Murphy, B. D., Rajkumar, K., Reyna, A. G. and Silversides, D. W. (1993b). Control of Luteal Function in the Mink (Mustela-Vison). *Fertility and Infertility in Dogs, Cats and Other Carnivores,* 47, 181-188.

Naeslund, G. (1979). The effect of glucose-, arginine- and leucine-deprivation on mouse blastocyst outgrowth in vitro. *Ups J Med Sci,* 84, 9-20.

Nieder, G. L. and Weitlauf, H. M. (1984). Regulation of glycolysis in the mouse blastocyst during delayed implantation. *J Exp Zool,* 231, 121-129.

Nilsson, B. O., Magnusson, C., Widehn, S. and Hillensjo, T. (1982). Correlation between blastocyst oxygen consumption and trophoblast cytochrome oxidase reaction at initiation of implantation of delayed mouse blastocysts. *J Embryol Exp Morphol,* 71, 75-82.

Noci, I., Fuzzi, B., Rizzo, R., Melchiorri, L., Criscuoli, L., Dabizzi, S., Biagiotti, R., Pellegrini, S., Menicucci, A. and Baricordi, O. R. (2005). Embryonic soluble HLA-G as a marker of developmental potential in embryos. *Hum Reprod,* 20, 138-146.

O'Neill, C. (2005). The role of paf in embryo physiology. *Hum Reprod Update,* 11, 215-228.

Papke, R. L., Concannon, P. W., Travis, H. F. and Hansel, W. (1980). Control of luteal function and implantation in the mink by prolactin. Journal of Animal Science, 50, 1102-1107.

Paria, B. C., Chakraborty, C. and Dey, S. K. (1990). Catechol estrogen formation in the mouse uterus and its role in implantation. *Mol Cell Endocrinol,* 69, 25-32.

Paria, B. C., Das, S. K., Andrews, G. K. and Dey, S. K. (1993a). Expression of the epidermal growth factor receptor gene is regulated in mouse blastocysts during delayed implantation. *Proceedings of the National Academy of Sciences of the United States of America,* 90, 55-59.

Paria, B. C., Huet-Hudson, Y. M. and Dey, S. K. (1993b). Blastocyst's state of activity determines the "window" of implantation in the receptive mouse uterus. *Proc Natl Acad Sci U S A,* 90, 10159-10162.

Paria, B. C., Das, S. K., Mead, R. A. and Dey, S. K. (1994). Expression of epidermal growth factor receptor in the preimplantation uterus and blastocyst of the western spotted skunk. *Biol Reprod,* 51, 205-213.

Paria, B. C., Das, S. K. and Dey, S. K. (1995). The preimplantation mouse embryo is a target for cannabinoid ligand-receptor signaling. *Proc Natl Acad Sci U S A,* 92, 9460-9464.

Paria, B. C., Deutsch, D. D. and Dey, S. K. (1996). The uterus is a potential site for anandamide synthesis and hydrolysis: differential profiles of anandamide synthase and hydrolase activities in the mouse uterus during the periimplantation period. *Mol Reprod Dev,* 45, 183-192.

Paria, B. C., Lim, H., Wang, X. N., Liehr, J., Das, S. K. and Dey, S. K. (1998). Coordination of differential effects of primary estrogen and catecholestrogen on two distinct targets mediates embryo implantation in the mouse. *Endocrinology,* 139, 5235-5246.

Paria, B. C., Ma, W., Tan, J., Raja, S., Das, S. K., Dey, S. K. and Hogan, B. L. (2001a). Cellular and molecular responses of the uterus to embryo implantation can be elicited by locally applied growth factors. *Proc Natl Acad Sci U S A,* 98, 1047-1052.

Paria, B. C., Song, H., Wang, X., Schmid, P. C., Krebsbach, R. J., Schmid, H. H., Bonner, T. I., Zimmer, A. and Dey, S. K. (2001b). Dysregulated cannabinoid signaling disrupts uterine receptivity for embryo implantation. *J Biol Chem,* 276, 20523-20528.

Passavant, C., Zhao, X., Das, S. K., Dey, S. K. and Mead, R. A. (2000). Changes in uterine expression of leukemia inhibitory factor receptor gene during pregnancy and its up-regulation by prolactin in the western spotted skunk. *Biol Reprod,* 63, 301-307.

Polejaeva, I. A., Reed, W. A., Bunch, T. D., Ellis, L. C. and White, K. L. (1997). Prolactin-induced termination of obligate diapause of mink (Mustela vison) blastocysts in vitro and subsequent establishment of embryonic stem-like cells. *J Reprod Fertil,* 109, 229-236.

Potts, D. M. (1968). The ultrastructure of implantation in the mouse. *J Anat,* 103, 77-90.

Ptak, G. E., Tacconi, E., Czernik, M., Toschi, P., Modlinski, J. A. and Loi, P. (2012). Embryonic diapause is conserved across mammals. *PLoS One,* 7, e33027.

Raab, G., Kover, K., Paria, B. C., Dey, S. K., Ezzell, R. M. and Klagsbrun, M. (1996). Mouse preimplantation blastocysts adhere to cells expressing the transmembrane form of heparin-binding EGF-like growth factor. *Development,* 122, 637-645.

Reese, J., Das, S. K., Paria, B. C., Lim, H., Song, H., Matsumoto, H., Knudtson, K. L., DuBois, R. N. and Dey, S. K. (2001). Global gene expression analysis to identify molecular markers of uterine receptivity and embryo implantation. *J Biol Chem,* 276, 44137-44145.

Renfree, M. B. and Tyndale-Biscoe, C. H. (1973). Intrauterine development after diapause in the marsupial Macropus eugenii. *Dev Biol,* 32, 28-40.

Renfree, M. B. and Shaw, G. (2000). Diapause. *Annu Rev Physiol,* 62, 353-375.

Richter, K. S., Harris, D. C., Daneshmand, S. T. and Shapiro, B. S. (2001). Quantitative grading of a human blastocyst: optimal inner cell mass size and shape. *Fertil Steril,* 76, 1157-1167.

Sakhuja, D., Sengupta, J. and Manchanda, S. K. (1982). A study of carbohydrate metabolism in 'delayed' and 'activated' mouse blastocyst and uterus. *J Endocrinol*, 95, 283-286.

Santos, F., Hyslop, L., Stojkovic, P., Leary, C., Murdoch, A., Reik, W., Stojkovic, M., Herbert, M. and Dean, W. (2010). Evaluation of epigenetic marks in human embryos derived from IVF and ICSI. *Hum Reprod*, 25, 2387-2395.

Scott, R., Seli, E., Miller, K., Sakkas, D., Scott, K. and Burns, D. H. (2008). Noninvasive metabolomic profiling of human embryo culture media using Raman spectroscopy predicts embryonic reproductive potential: a prospective blinded pilot study. *Fertil Steril*, 90, 77-83.

Seli, E., Sakkas, D., Scott, R., Kwok, S. C., Rosendahl, S. M. and Burns, D. H. (2007). Noninvasive metabolomic profiling of embryo culture media using Raman and near-infrared spectroscopy correlates with reproductive potential of embryos in women undergoing in vitro fertilization. *Fertil Steril*, 88, 1350-1357.

Sher, G., Keskintepe, L., Fisch, J. D., Acacio, B. A., Ahlering, P., Batzofin, J. and Ginsburg, M. (2005). Soluble human leukocyte antigen G expression in phase I culture media at 46 hours after fertilization predicts pregnancy and implantation from day 3 embryo transfer. *Fertil Steril*, 83, 1410-1413.

Sherwin, J. R., Freeman, T. C., Stephens, R. J., Kimber, S., Smith, A. G., Chambers, I., Smith, S. K. and Sharkey, A. M. (2004). Identification of genes regulated by leukemia-inhibitory factor in the mouse uterus at the time of implantation. *Mol Endocrinol*, 18, 2185-2195.

Song, H., Lim, H., Paria, B. C., Matsumoto, H., Swift, L. L., Morrow, J., Bonventre, J. V. and Dey, S. K. (2002). Cytosolic phospholipase A2alpha is crucial [correction of A2alpha deficiency is crucial] for 'on-time' embryo implantation that directs subsequent development. *Development*, 129, 2879-2889.

Song, J. H., Houde, A. and Murphy, B. D. (1998). Cloning of leukemia inhibitory factor (LIF) and its expression in the uterus during embryonic diapause and implantation in the mink (Mustela vison). *Mol Reprod Dev*, 51, 13-21.

Spindler, R. E., Renfree, M. B. and Gardner, D. K. (1996). Carbohydrate uptake by quiescent and reactivated mouse blastocysts. *J Exp Zool*, 276, 132-137.

Stachecki, J. J. and Armant, D. R. (1996). Transient release of calcium from inositol 1,4,5-trisphosphate-specific stores regulates mouse preimplantation development. *Development*, 122, 2485-2496.

Stavreus-Evers, A., Aghajanova, L., Brismar, H., Eriksson, H., Landgren, B. M. and Hovatta, O. (2002). Co-existence of heparin-binding epidermal growth factor-like growth factor and pinopodes in human endometrium at the time of implantation. *Mol Hum Reprod,* 8, 765-769.

Stoufflet, I., Mondain-Monval, M., Simon, P. and Martinet, L. (1989). Patterns of plasma progesterone, androgen and oestrogen concentrations and in-vitro ovarian steroidogenesis during embryonic diapause and implantation in the mink (Mustela vison). *J Reprod Fertil,* 87, 209-221.

Sturmey, R. G., Hawkhead, J. A., Barker, E. A. and Leese, H. J. (2009). DNA damage and metabolic activity in the preimplantation embryo. *Hum Reprod,* 24, 81-91.

Tachi, S., Tachi, C. and Lindner, H. R. (1970). Ultrastructural features of blastocyst attachment and trophoblastic invasion in the rat. *J Reprod Fertil,* 21, 37-56.

Tang, F., Barbacioru, C., Wang, Y., Nordman, E., Lee, C., Xu, N., Wang, X., Bodeau, J., Tuch, B. B., Siddiqui, A., Lao, K. and Surani, M. A. (2009). mRNA-Seq whole-transcriptome analysis of a single cell. *Nat Methods,* 6, 377-382.

Threadgill, D. W., Dlugosz, A. A., Hansen, L. A., Tennenbaum, T., Lichti, U., Yee, D., LaMantia, C., Mourton, T., Herrup, K., Harris, R. C. and et al. (1995). Targeted disruption of mouse EGF receptor: effect of genetic background on mutant phenotype. *Science,* 269, 230-234.

Torbit, C. A. and Weitlauf, H. M. (1974). The effect of oestrogen and progesterone on CO_2 production by 'delayed implanting' mouse embryos. *J Reprod Fertil,* 39, 379-382.

Tranguch, S., Daikoku, T., Guo, Y., Wang, H. and Dey, S. K. (2005). Molecular complexity in establishing uterine receptivity and implantation. *Cell Mol Life Sci,* 62, 1964-1973.

Turner, K., Martin, K. L., Woodward, B. J., Lenton, E. A. and Leese, H. J. (1994). Comparison of pyruvate uptake by embryos derived from conception and non-conception natural cycles. *Hum Reprod,* 9, 2362-2366.

Urbanski, J. P., Johnson, M. T., Craig, D. D., Potter, D. L., Gardner, D. K. and Thorsen, T. (2008). Noninvasive metabolic profiling using microfluidics for analysis of single preimplantation embryos. *Anal Chem,* 80, 6500-6507.

van Amerongen, R. and Nusse, R. (2009). Towards an integrated view of Wnt signaling in development. *Development,* 136, 3205-3214.

Van Blerkom, J., Chavez, D. J. and Bell, H. (1978). Molecular and cellular aspects of facultative delayed implantation in the mouse. *Ciba Found Symp,* 141-172.

Van Soom, A., Vanroose, G. and de Kruif, A. (2001). Blastocyst evaluation by means of differential staining: a practical approach. *Reprod Domest Anim,* 36, 29-35.

Wakefield, S. L., Lane, M. and Mitchell, M. (2011). Impaired mitochondrial function in the preimplantation embryo perturbs fetal and placental development in the mouse. *Biol Reprod,* 84, 572-580.

Wang, H., Matsumoto, H., Guo, Y., Paria, B. C., Roberts, R. L. and Dey, S. K. (2003). Differential G protein-coupled cannabinoid receptor signaling by anandamide directs blastocyst activation for implantation. *Proc Natl Acad Sci U S A,* 100, 14914-14919.

Wang, H., Guo, Y., Wang, D., Kingsley, P. J., Marnett, L. J., Das, S. K., DuBois, R. N. and Dey, S. K. (2004a). Aberrant cannabinoid signaling impairs oviductal transport of embryos. *Nat Med,* 10, 1074-1080.

Wang, H. and Dey, S. K. (2006). Roadmap to embryo implantation: clues from mouse models. *Nat Rev Genet,* 7, 185-199.

Wang, H., Dey, S. K. and Maccarrone, M. (2006). Jekyll and hyde: two faces of cannabinoid signaling in male and female fertility. *Endocr Rev,* 27, 427-448.

Wang, J., Mayernik, L., Schultz, J. F. and Armant, D. R. (2000). Acceleration of trophoblast differentiation by heparin-binding EGF-like growth factor is dependent on the stage-specific activation of calcium influx by ErbB receptors in developing mouse blastocysts. *Development,* 127, 33-44.

Wang, J., Mayernik, L. and Armant, D. R. (2002). Integrin signaling regulates blastocyst adhesion to fibronectin at implantation: intracellular calcium transients and vesicle trafficking in primary trophoblast cells. *Dev Biol,* 245, 270-279.

Wang, Q. T., Piotrowska, K., Ciemerych, M. A., Milenkovic, L., Scott, M. P., Davis, R. W. and Zernicka-Goetz, M. (2004b). A genome-wide study of gene activity reveals developmental signaling pathways in the preimplantation mouse embryo. *Dev Cell,* 6, 133-144.

Warner, C. M., Newmark, J. A., Comiskey, M., De Fazio, S. R., O'Malley, D. M., Rajadhyaksha, M., Townsend, D. J., McKnight, S., Roysam, B., Dwyer, P. J. and DiMarzio, C. A. (2004). Genetics and imaging to assess oocyte and preimplantation embryo health. *Reprod Fertil Dev,* 16, 729-741.

Weitlauf, H., Kiessling, A. and Buschman, R. (1979). Comparison of DNA polymerase activity and cell division in normal and delayed-implanting mouse embryos. *J Exp Zool,* 209, 467-472.

Weitlauf, H. M. and Greenwald, G. S. (1968). Influence of estrogen and progesterone on the incorporation of 35S methionine by blastocysts in ovariectomized mice. *J Exp Zool,* 169, 463-469.

Weitlauf, H. M. (1973a). Changes in the protein content of blastocysts from normal and delayed implanting mice. *Anat Rec,* 176, 121-123.

Weitlauf, H. M. (1973b). In vitro uptake and incorporation of amino acids by blastocysts from intact and ovariectomized mice. *J Exp Zool,* 183, 303-308.

Wilcox, A. J., Baird, D. D. and Weinberg, C. R. (1999). Time of implantation of the conceptus and loss of pregnancy. *N Engl J Med,* 340, 1796-1799.

Wu, J. T. and Meyer, R. K. (1974). Ultrastructural changes of rat blastocysts induced by estrogen during delayed implantation. *Anat Rec,* 179, 253-272.

Xie, H., Wang, H., Tranguch, S., Iwamoto, R., Mekada, E., Demayo, F. J., Lydon, J. P., Das, S. K. and Dey, S. K. (2007). Maternal heparin-binding-EGF deficiency limits pregnancy success in mice. *Proc Natl Acad Sci U S A,* 104, 18315-18320.

Xie, H., Tranguch, S., Jia, X., Zhang, H., Das, S. K., Dey, S. K., Kuo, C. J. and Wang, H. (2008). Inactivation of nuclear Wnt-beta-catenin signaling limits blastocyst competency for implantation. *Development,* 135, 717-727.

Yang, Z. M., Paria, B. C. and Dey, S. K. (1996). Activation of brain-type cannabinoid receptors interferes with preimplantation mouse embryo development. *Biol Reprod,* 55, 756-761.

Ye, X., Hama, K., Contos, J. J., Anliker, B., Inoue, A., Skinner, M. K., Suzuki, H., Amano, T., Kennedy, G., Arai, H., Aoki, J. and Chun, J. (2005). LPA3-mediated lysophosphatidic acid signalling in embryo implantation and spacing. *Nature,* 435, 104-108.

Yoo, H. J., Barlow, D. H. and Mardon, H. J. (1997). Temporal and spatial regulation of expression of heparin-binding epidermal growth factor-like growth factor in the human endometrium: a possible role in blastocyst implantation. *Dev Genet,* 21, 102-108.

Yoshinaga, K. and Adams, C. E. (1966). Delayed implantation in the spayed, progesterone treated adult mouse. *J Reprod Fertil,* 12, 593-595.

Yoshinaga, K. (1980). Inhibition of implantation by advancement of uterine sensitivity and refractoriness. *Blastocysts-endometrium relationships:*

progress in reproductive biology. Leroy F., Finn CA, Psychoyos A., Hubinot PO (eds.), Karger, Basel, Switzerland, 189-199.

Yu, Y., Wu, J., Fan, Y., Lv, Z., Guo, X., Zhao, C., Zhou, R., Zhang, Z., Wang, F., Xiao, M., Chen, L., Zhu, H., Chen, W., Lin, M., Liu, J., Zhou, Z., Wang, L., Huo, R., Zhou, Q. and Sha, J. (2009). Evaluation of blastomere biopsy using a mouse model indicates the potential high risk of neurodegenerative disorders in the offspring. *Mol Cell Proteomics,* 8, 1490-1500.

Zhou, W., Choi, M., Margineantu, D., Margaretha, L., Hesson, J., Cavanaugh, C., Blau, C. A., Horwitz, M. S., Hockenbery, D., Ware, C. and Ruohola-Baker, H. (2012). HIF1alpha induced switch from bivalent to exclusively glycolytic metabolism during ESC-to-EpiSC/hESC transition. *EMBO J,* 31, 2103-2116.

In: Embryo Development
Editors: D. Reyes and A. Casales

ISBN: 978-1-62417-723-1
© 2013 Nova Science Publishers, Inc.

Chapter 3

RECENT ADVANCES IN THE STUDY OF LIMB DEVELOPMENT: THE EMERGENCE AND FUNCTION OF THE APICAL ECTODERMAL RIDGE

Joaquin Rodriguez-Leon[1,2,*], *Ana Raquel Tomas*[1,2], *Austin Johnson*[3] *and Yasuhiko Kawakami*[3,4,5]

[1]Department de Anatomía Humana, Biología Celular y Zoología, Facultad de Medicina, Universidad de Extremadura, Badajoz, Spain
[2]Instituto Gulbenkian de Ciência, Portugal
[3]Department of Genetics, Cell Biology and Development,
[4]Stem Cell Institute,
[5]Developmental Biology Center, University of Minnesota, Minneapolis, MN, US

ABSTRACT

Vertebrate extremities develop from limb buds, which emerge as paired protrusions in the lateral plate mesoderm. Forelimb buds are located anteriorly and hindlimb buds are positioned posteriorly. The morphogenesis of the limb requires coordinated actions of several

---

[*] Corresponding authors: Joaquin Rodriguez-Leon, Email: jrleon@unex.es, Yasuhiko Kawakami, Email: kawak005@umn.edu.

organizing centers, among which the apical ectodermal ridge (AER) plays crucial roles in limb development. Recent studies have shown how the life of the AER (induction, maturation, maintenance and regression) is regulated.

This regulation includes cell type- and process- specific roles of previously identified molecules, such as fibroblast growth factors (FGFs), Wnts and bone morphogenetic proteins (BMPs). The studies have also revealed several new players, such as Arid3b, R-Spondin 2 and Flrt3. These advances have enhanced the understanding of how the AER is regulated from its emergence to its regression. Progress has also been made in understanding AER function in relation to processes critical for limb development: proximal-distal patterning, anterior-posterior patterning, chondrogenesis and apoptosis. By focusing on two major model systems, chick and mouse embryos, we will review recent advances in combination with relevant previous studies in the development and function of the AER.

INTRODUCTION

Vertebrate limb development is an excellent model system for the study of pattern formation and morphogenesis. The processes of vertebrate limb development requires the coordination of numerous cellular and molecular mechanisms, resulting in a group of cells being transformed into the many different types of extremities seen in present-day vertebrates. A number of excellent review papers published to date describe the mechanisms and processes of vertebrate limb development [1-9].

In this review, we aim to focus on recent progress in the study of the development and function of the apical ectodermal ridge (AER), a specialized, transiently-present, ectodermal structure, which regulates outgrowth and pattern formation of the limb. We will focus on reports on chick and mouse embryos, two major vertebrate models whose anatomical features of the limb are similar to those in humans. To avoid redundancy with previous publications, first, we shortly summarize the processes of early stages of limb development with recent reports.

We then describe the development of the AER and studies of the functions of the AER. In the concluding paragraph, we discuss how the study of limb development can contribute to enhancing our knowledge of biological and biomedical sciences by highlighting a few specific research topics.

Recent Advances in the Study of Limb Development

EMERGENCE OF THE LIMB BUD

The development of the limb provides an experimental platform to understand a variety of biological processes. These include not only patterning and morphogenesis, but also positioning in the developing embryo and initiation of the development, maintenance and termination of outgrowth, and differentiation. It also serves as a system to study how diverse morphologies are generated, while common skeletal features are conserved among species, as reviewed in [10, 11], as well as how two types of limbs, namely the forelimbs and hindlimbs, generate their differences.

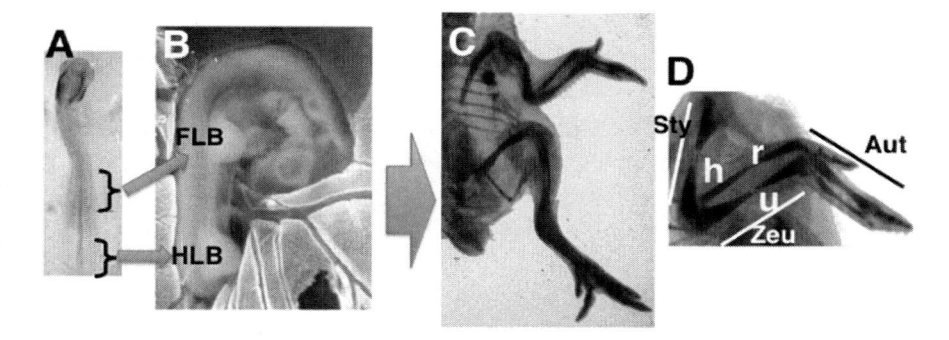

Figure 1. Limb development in the chick embryos. (A): Image of 2.5-day chick embryo. Brackets indicate forelimb- and hindlimb-bud forming region. (B): Image of 3.5-day chick embryo. The forelimb bud (FLB) and hindlimb bud (HLB) are established. (C): Image of cartilage-stained 10-day chick embryo. (D): Close up image of the forelimb skeleton of 10-day chick embryo. The three regions along the proximal-distal axis are indicated as the stylopod (sty), zeugopod (zeu) and autopod (aut) regions. h, r and u represents humerus in the stylopod region, and the radius and ulna in the zeugopod region, respectively.

Limb development begins with the formation of two pairs of limb buds within defined areas of the developing embryo (Figure 1). During body extension, cells in the broad region of the lateral plate mesoderm (LPM) proliferate [12]. This classic study demonstrated that cells at the limb-forming region continue to proliferate, while cells in the flank reduce their rate of proliferation. Recent reports have found that early morphogenesis of the simple limb bud structure involves not only differential proliferation between the limb-forming and surrounding areas but also cell-cell interaction and cell migration. For instance, it has been found that the limb bud mesoderm is more cohesive than the surrounding flank mesoderm, suggesting that budding may actually be caused by the limb mesoderm's higher liquid-like cohesivity [13]. Moreover, directed cell mobility toward the area of outgrowth from the surrounding LPM, and directed division in and around the limb bud-forming region play a role in the shaping of the limb bud at the stage of initial budding [14]. Molecular studies suggest that WNT5a, a secreted factor of the Wnt family, acts as an attractant to regulate mesoderm cell migration during the process of oriented cell migration toward the limb bud [14, 15].

THE DEVELOPMENT OF THE AER

After initial limb budding, the AER is formed at the dorsal-ventral boundary at the distal edge of the limb bud (Figure 2). One of the characteristic features of the AER is the multilayered ectodermal structure. The AER is a transient structure whose life can be divided into four stages: initiation, maturation, maintenance and regression. These stages, well studied both in chick and mouse embryos, involve ectodermal cells that change their shape and gene expression throughout the process to contribute to the AER [16, 17].

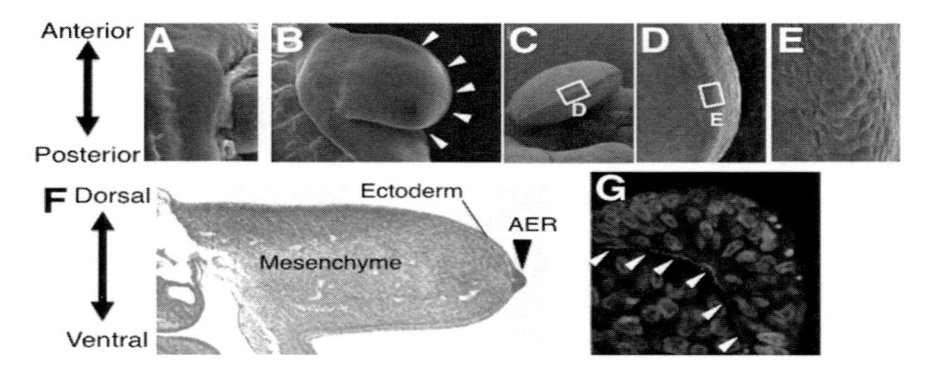

Figure 2. Morphology of the AER in the developing limb. (A-E): Scanning electron micrograms of chick forelimb bud at 2.75-day (A) and 4-day (B-E). The AER is not formed in the 2.75-day embryo, but, is clearly visible in the 4-day embryo (arrowheads in B). D and E show close up of the boxed areas in C and D, respectively. (F): Cross section of the forelimb bud of 3-day chick embryo. The AER, denoted with an arrowhead, is located at the dorsal-ventral boundary. (G): Confocal image of the distal region of the forelimb bud of a 3-day chick embryo. Arrowheads point to laminin signal at the border of the mesenchyme and ectoderm tissue. White oval signals show nuclear DAPI signals. The AER can be observed as a pseudostratified ectodermal structure.

Initiation of the AER Development

Prior to the formation of the AER, ectodermal cells in the limb bud-forming region begin to express fibroblast growth factor 8 (*Fgf8*), a member of the FGF family of secreted factors (Figure 3). These *Fgf8*-expressing cells are considered to be AER precursor cells and are located broadly in the limb bud-forming region and in a patchy pattern, referred to as the pre-AER. Interestingly, AER precursor cells are located in a broad region, in both dorsal and ventral ectoderm, in chick embryos [16], while they are located only in the ventral ectoderm in the mouse embryo [17].

Studies of secreted molecules for cell-cell signaling, including FGF, Wnt and bone morphogenetic protein (BMP), have shown that initiation of AER formation requires FGF10, another member of the FGF family emanating from the mesenchyme of the limb bud-forming region [18, 19]. In the absence of *Fgf10*, *Fgf8* is not induced in the ectoderm, indicating that AER precursor cells were not specified. Evidence obtained in both chick and mouse embryos demonstrated that ectodermal Wnt/ß-catenin signaling, which is regulated by mesenchymal FGF10, induces ectodermal *Fgf8* [20-23]. During this process, AER precursor cells require FGFR2 to respond to FGF10, since mice lacking an isoform of the FGFR2b (an isoform of the receptor specifically expressed in the ectoderm of the limb bud) fail to develop the AER [24, 25]. Thus, a signal cascade of mesenchymal FGF10 -> ectodermal FGFR2b -> Wnt/ß-catenin -> *Fgf8* expression operates to induce *Fgf8* expression in the AER precursor cells.

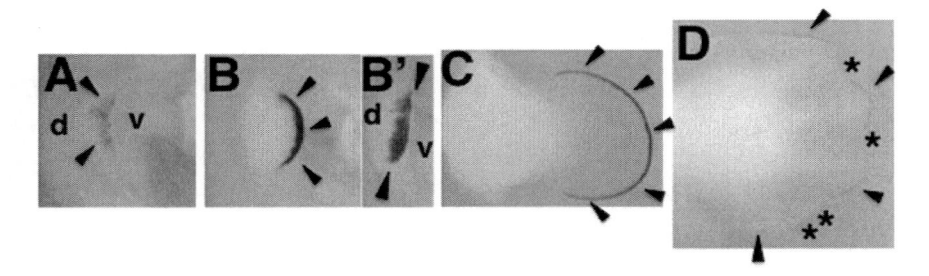

Figure 3. *Fgf8* expression during limb development. (A-D): Whole mount in situ hybridization images of *Fgf8* expression in the forelimb buds of the mouse embryo. *Fgf8* is expressed in the AER precursor cells at E9.0 (A), and in the AER at E10.5 (B, B'). *Fgf8* expression expands as the limb bud grows at E11.5 (C), but is downregulated in the E12.5 limb bud (D). Arrowheads point to *Fgf8* expression, and asterisks in D indicates the region without detectable *Fgf8*. A and B' are distal views, marked with dorsal (d) and ventral (v) sides. B, C and D are dorsal views.

Moreover, the early inactivation of the BMP receptor 1a (*Bmpr1a*) gene in the ectoderm of the limb bud-forming region resulted in the failure to initiate *Fgf8* expression, indicating that ectodermal BMP signaling is also required for initiating AER formation [26]. An elegant genetic study that combined both loss- and gain- of function of Wnt and BMP signaling showed that ectodermal BMP signaling acts upstream of Wnt/ß-catenin signaling [23, 27]. It has not yet been determined whether Wnt and BMP in the ectoderm act in a non-cell autonomous manner to induce the patchy pattern of *Fgf8*-expressing AER precursor cells. Given that the mechanism of *Fgf8* induction in the AER precursor cells is unknown, the mechanisms that specify AER precursor cells in the ectoderm remain unidentified.

AER Maturation

During the maturation step, AER precursor cells migrate toward, and intercalate at, the dorsal-ventral boundary (a process known as "apical compaction"), forming a morphologically-distinguishable, multi-layered structure [16, 17] (Figure 2). Correlating with the positioning of the AER at the dorsal-ventral boundary, the maturation process is regulated by proper dorsal-ventral patterning. A key factor for the establishment of the dorsal-ventral boundary is *En1*, which is one of two homeobox genes related to *Drosophila engrailed*, and is expressed in the ventral ectoderm of the limb bud [28]. *En1* restricts the expression of *Wnt7a* to the dorsal ectoderm. When the dorsal-ventral boundary in the ectoderm is disturbed by either loss- or gain- of *En1* function, the formation of the AER is also disrupted [29, 30]. Thus, the expression of *En1* in the ventral ectoderm defines the dorsal-ventral boundary of the limb ectoderm, which is essential for AER maturation. Furthermore, BMP signaling acts upstream of *En1* since forced activation of BMP signaling in the limb ectoderm in the chick embryo could induce *En1* expression in the dorsal ectoderm [31]. Conversely, conditionally inactivating *Bmpr1a* caused the loss of *En1* expression in the limb bud ectoderm [27]. Three *Bmp* ligand genes, *Bmp2*, *Bmp4* and *Bmp7*, are expressed in the limb bud ectoderm and mesenchyme [32]. Due to functional redundancy among them, functional testing of the role of BMP signaling was conducted by using an antagonist. Inhibiting BMP activities by transgenic expression of Noggin [33, 34], a BMP antagonist, in the AER caused the loss of *En1* expression and ectopic expression of *Wnt7a* in the ventral ectoderm, leading to disruption of the dorsal-ventral boundary. This was correlated with a broad *Fgf8*-expression domain along the dorsal-ventral axis in the ectoderm, with an autopod skeleton exhibiting polydactyly and syndactyly. More recently, two *Bmp* genes, *Bmp2* and *Bmp4*, were specifically inactivated in the AER by a conditional knockout approach. This resulted in the loss of *En1* expression and increased *Fgf* expression levels in the AER [35]. The resulting phenotype displayed syndactyly and polydactyly, similar to the *Noggin* transgenic mice [33]. The fact that *Bmp7* function is not disrupted in the *Bmp2-Bmp4* double mutants [35] and the reported limb agenesis in AER-specific *Bmpr1a* inactivation [27] suggest that BMPs also have a role in limb outgrowth. Nonetheless, these reports highlight the BMP-*En1* pathway as an important player in AER maturation (Figure 4B). In addition to the BMP-*En1* pathway, three transcription factors, p63, Sp8 and Arid3b, have been shown to be essential for AER maturation rather than induction. p63, a homolog of the tumor-suppressor p53, is expressed broadly in embryonic ectodermal cells, including the ectoderm of the limb bud-forming region. *Fgf8* expression is induced in the ectoderm in $p63^{-/-}$ embryos, indicating that the AER precursor cells are specified. However, the limb bud ectoderm cells in $p63^{-/-}$ embryos do not form the multilayered structure, and fail to develop into the AER [36, 37]. The defect of AER maturation in the absence of *p63* seems to involve abnormal dorsal-ventral patterning [36, 37]. However, reduced *Fgf8* expression, as well as the failure to maintain the expression of genes involved in normal limb outgrowth in *p63* mutant limbs, suggest that *p63* is also involved in the maintenance of ectodermal cells characteristics. In contrast, functions of *Sp8* and *Arid3b* do not appear to involve dorsal-ventral patterning of the limb ectoderm, given normal expression of limb ectoderm genes in these mutants [38-40]. *Sp8*, which is structurally and functionally related to the *Drosophila buttonhead* gene [39, 41], is a member of the Sp-KLF family of zinc finger transcription factors. In $Sp8^{-/-}$ embryos, AER precursor cells are induced, but fail to form morphologically distinguishable AER [38, 39], similar to $p63^{-/-}$ embryos. A more recent analysis demonstrated that *Sp6*, another member of the Sp-KLF family, is also required for the AER maturation process [42]. Loss of *Sp6* resulted in the failure to complete the maturation process, producing an AER that was broad and flat. It is likely that functional redundancy exists between *Sp6* and *Sp8*, and thus, it would be interesting to examine in the future whether a double mutant of *Sp6* and *Sp8* shows other phenotypes, such as defects in the induction of AER precursor cells. *Arid3b* is one of three genes of the *Arid3* group, which is homologous to the *Drosophila dead ringer* gene. Loss of *Arid3b* function in chick and mouse embryos caused defective AER maturation (40).

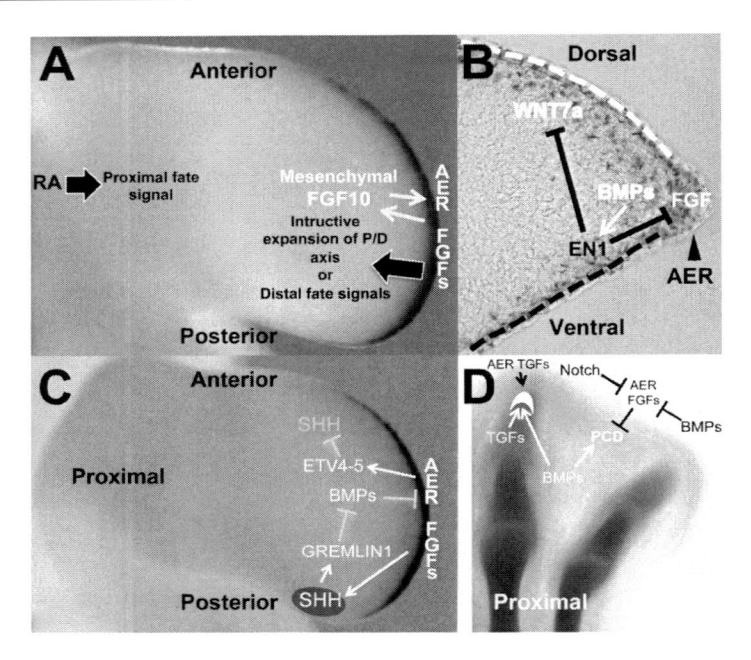

Figure 4. Functions of the AER during limb development. (A):Proximal-distal patterning: Forelimb bud of a 4-day chick embryo, labeled by *Fgf19* expression in the AER. In white, the consensus reciprocal loop between FGF10 from the mesenchyme and FGFs from the AER is represented. Black arrows in the distal and proximal regions represent proposed activities of AER in the distal mesenchyme and a putative activity of retinoic acid, respectively, in the two signal gradient model. (B): Relationship of AER with D/V signals: Transverse section of a limb bud from a 3-day chick embryo. White and black dashed lines represent expression of *Wnt7a* and *En1* in the dorsal and ventral ectoderm, respectively. BMP activity, from mesenchyme and AER, positively regulates *En1* expression, which, in turn, negatively controls *Wnt7a* expression in the dorsal ectoderm and *Fgf* expression in the AER. (C): AER and the anterior-posterior patterning, which involves the SHH-GREMLIN1-FGF loop: Forelimb bud of a 5-day chick embryo, labeled by *Fgf19* expression in the AER. FGFs from AER positively regulate *Shh* expression in the ZPA (grey oval), and inhibit *Shh* expression in the anterior limb bud through *Etv4* and *Etv5* genes in the mesenchyme. SHH activates *Gremlin1* expression in the mesenchyme, which protects AER from the negative action of BMPs. (D): AER activity and chondrogenic and apoptotic signals: Alcian green stained hindlimb from a 7-day chick embryo where digits III, IV and the third interdigital space are magnified. Notch signaling and BMP signaling in the AER negatively regulate AER size and the level of FGF signaling. The balance between FGF signaling from the AER and apoptotic signals controls programmed cell death in the interdigital area. AER activity from the AER is needed to maintain TGFß activity at the tip of forming digits (white crescent) allowing chondrogenic signals from the mesenchyme to trigger cartilage formation. RA, Retinoic Acid; PCD, programmed cell death.

In this experiment, the AER precursor cells exhibited normal gene expression and proliferation, but an increase in stress fibers and aberrant distribution of actin filaments. Further analysis *in vitro* using *Arid3b* mutant embryonic fibroblasts, as well as cell tracing *in vivo*, demonstrated that *Arid3* was required for cell motility and rearrangements.

Given that *p63*, *Sp8*, *Sp6* and *Arid3b* encode transcription factors, their targets could be involved in the cell motility and adhesion of the AER precursor cells that are necessary for migration to the dorsal-ventral boundary and subsequent formation of the multi-layered AER structure. Although specific factors downstream of the transcription factors are still unknown, it is interesting to note that p63 has been shown to regulate an adhesion program, and that its targets include ICAM3, integrins (such as ß1, ß4, α1), fibronectin and P-cadherin [43-47]. Thus, changes in cell adhesion and motility might be involved in the maturation of the AER. Similar to WNT5a-dependent mesenchymal cell migration during the initial outgrowth process of the limb bud, factors expressed at the distal edge of the dorsal-ventral boundary might serve as an attractant for the AER precursor cells during the maturation process.

AER Maintenance

The maintenance of the AER involves interactions with the underlying mesenchyme. *Fgf10*, expressed in the distal mesenchyme, and *Sonic hedgehog* (*Shh*), expressed in the posterior margin of the mesenchyme, defined as the zone of polarizing activity (ZPA), are key factors in the interaction between the AER and mesenchyme (Figure 4A and C). *Bmp2, Bmp4* and *Bmp7,* as well as *Gremlin1*, which encodes a BMP antagonist, are critical components of the mesenchymal SHH-GREMLIN1 – AER-FGF feedback loop [48-53]. In this loop, GREMLIN1 inhibits the function of BMPs, which are expressed in the ectoderm and mesenchyme [32] and downregulate *Fgf* expression in the AER [31, 33, 34]. Another feedback loop by the mesenchymal FGF10 – AER-FGF8 [18, 19, 25, 54] also maintains the AER, which will be further discussed later in relation to the function of the AER. Rather than re-emphasizing the role of these signaling loops, we describe signaling within the AER required for AER maintenance. Recent studies have highlighted roles of ß-catenin signaling, FGF signaling and hedgehog signaling within the AER.

Activation of ß-catenin signaling in the AER has been demonstrated by expression of target genes, such as *Axin2* [55], as well as synthetic reporter genes, such as Top-Gal and BATgal [56-58]. The expression pattern of Wnt ligands differs in chick and mouse embryos. In particular, *Wnt3a* is expressed in AER precursors and in mature AER cells in the chick [20, 21]. In the mouse, *Wnt3a* is not expressed in the limb bud, but *Wnt3* is broadly expressed in the limb ectoderm [22]. These Wnt ligands activate ß-catenin signaling for the AER induction (described above) and maintenance. However, involvement of a non-Wnt ligand for the activation of ß-catenin signaling has become evident. R-spondin is a secreted molecule and functions to activate ß-catenin signaling upstream of the Frizzled proteins, which are transmembrane proteins that function as components of the Wnt receptor complex. Several studies have shown that *R-spondin2* is expressed in the AER [59-62], and loss of *R-spondin2* caused failure to maintain the AER in the mouse embryos [60-62]. These studies indicate that, in addition to the well-established role of Wnt/ß-catenin signaling in the maintenance of the AER, *R-spondin2* also participates in the activation of ß-catenin signaling for AER maintenance.

Within the AER, FGF signaling through FGFR2 functions to maintain the AER, similar to the initiation process. AER-specific loss of *Fgfr2* using the *Msx2-cre* transgene and the conditional allele of *Fgfr2*, caused failure to maintain both the multi-layered structure of the AER as well as *Fgf8* expression after the formation of the AER [63]. The importance of FGF signaling in the AER [64] is also demonstrated by studies of the transmembrane protein, Fibronection-Leucine-Rich-Transmembrane 3 (Flrt3), which interacts with FGFR [65]. Flrt3 is implicated in the activation of the intracellular FGF signaling pathway [66, 67], although mouse embryos lacking the *Flrt3* function exhibited normal FGF signaling in the posterior part of the body at early somitogenesis stages (E8.0 – 8.5) [68]. In the chick limb bud, overexpression of *Flrt3* has been shown to result in enlargements of pre-existing AER and, conversely, silencing *Flrt3* caused loss of the integrity of the AER [65].

These studies on ß-catenin signaling and FGF signaling indicate that factors, which are not considered as classical ligands or receptors, participate in the regulation of signaling in the AER during its maintenance.

Interestingly, activation of ß-catenin signaling concomitantly with conditional inactivation of *Fgfr2* in the mouse AER could rescue maintenance of the morphological AER as well as *Fgf8* expression [63]. This suggests that ß-catenin signaling functions downstream of FGF signaling for AER maintenance. The observation that Wnt/ß-catenin signaling could induce *Flrt3* suggests that a feedback mechanism regulates cross-talk between FGF signaling and ß-catenin signaling in the AER [65].

Activation of hedgehog signaling within the AER has been recently demonstrated. Mesenchymal SHH has been shown to regulate *Fgf* expression in the AER through Gremlin1-mediated inhibition of BMP signaling [50, 53]. It was thought that Shh signaling is not present in the ectoderm, including the AER [69-71]. However, gene expression analyses in the developing limb bud suggested activation of Shh signaling in the AER [72, 73]. Furthermore, SHH protein was detected in the AER [73]. Experimental manipulation of the levels of Shh signaling in the AER caused an alteration in the AER length. For instance, increasing the levels of Shh signaling by Shh-treated bead implantation and genetically abrogating Shh signaling in the AER caused shorter and longer AERs than normal limb buds [73]. Rather than simply maintaining the AER, Shh signaling-dependent regulation of the length of AER is shown to affect the posterior mesenchyme by altered levels of FGF from the AER. Due to the close proximity of AER and *Shh*-expression domains [48, 74-76], this mechanism, in turn, regulates *Shh* expression in the posterior mesenchyme to ensure proper levels of *Shh* expression through the SHH-GREMLIN1-FGF feedback loop [73]. However, continued activation of Shh signaling causes severe and pleiotropic patterning defects, observed as a polydactyly, polysyndactyly and oligodactyly [77]. Thus, maintenance of the AER during normal limb development is dependent upon a functional SHH-GREMLIN1-FGF regulatory loop and proper levels of SHH signaling within the AER.

Regression of the AER

Terminating growth of the limb is equally as important as initiating and maintaining limb outgrowth for size regulation. Since the AER maintains limb outgrowth, termination of limb growth is achieved by regression of the AER, which is initiated by disconnecting the SHH-GREMLIN1-FGF feedback loop required for AER maintenance. One of the disconnecting mechanisms involves tissue expansion in the limb. Analysis in chick embryos shows that *Shh*-expressing cells and their descendants do not express *Gremlin1*. As these cells expand, the space between *Gremlin1*-expressing cells and *Shh*-expressing cells is also enlarged, leading to failure to maintain *Gremlin1* expression by SHH. This disrupts the SHH-GREMLIN1-FGF feedback loop, leading to regression of the AER and termination of limb outgrowth [78]. A study in the mouse embryo showed a similar, tissue growth-dependent termination of the feedback loop [52, 79]. This analysis, however, uncovered a novel inhibitory role of

AER-derived FGF to repress *Gremlin1* expression. In this model, low levels of AER-FGFs participate in the SHH-GREMLIN1-FGF loop to maintain the AER. As FGF signaling elevates during the development of the limb bud, however, the level of FGF signaling reaches a threshold, above which it inhibits *Gremlin1* expression. This causes mesenchymal cells under the AER to fail to express *Gremlin1*, and regression of *Gremlin1*-expressing domain causes a disconnection in the SHH-GREMLIN1-FGF loop. Although timing of gene expression and details of the proposed mechanisms differ in these two models, it is of interest to note that tissue growth, which is one of the functions of the AER, causes termination of the signaling loop and the AER itself to regress.

As summarized, recent studies add new players in our understanding of the AER emergence, maturation, maintenance and regression. In the following section, we review classic and recent studies of the functions and roles of the AER during limb development.

FUNCTIONS OF THE AER

The AER functions in proper limb development. In this section, we review recent studies on the function of the AER for specific aspects of limb development, such as proximal-distal patterning, anterior-posterior patterning, chondrogenesis and apoptosis. The contributions of the AER to these processes initially came from classical manipulation experiments in the chick embryo, and have recently been studied in more detail using molecular and genetic tools in both chick and mouse embryos.

Role of AER in Proximal-Distal Patterning

Although the structure of the AER differs among vertebrates, the functional importance of the AER for limb outgrowth and patterning is a characteristic and conserved feature during vertebrate limb development. Classical experiments using chick embryos have shown that surgical removal of the AER results in cell death in the mesenchyme, resulting in truncation of the limb skeleton [80-82]. It is clear now that the feedback loop between the AER and the underlying mesenchyme, which is mediated by FGF signaling, is key to the prevention of mesenchymal cell death and maintenance of limb outgrowth. However, how the AER regulates pattern formation along the proximal-distal axis in relation to limb outgrowth remains controversial. The AER produces multiple FGFs, including FGF4, FGF8, FGF9 and FGF17 in the mouse embryo,, which are termed as the AER-FGFs [83, 84]. In the chick embryo, *Fgf19*, instead of *Fgf17* is expressed in the AER [85, 86]. Expression analysis in the mouse embryo demonstrated that only *Fgf8* is expressed throughout the AER, and *Fgf8* expression precedes that of other AER-*Fgfs* [87-96]. Elegant genetic analyses in mice demonstrated that FGF8 is the most important AER-FGF for limb development, although other FGFs can partially compensate for the function of FGF8 [84, 97]. AER-FGFs signal to the mesenchyme beneath the AER and mesenchymal cells express *Fgf10* in response to AER-FGFs, which, in turn, maintains *Fgf8* expression in the AER [18, 19, 25, 54] (Figure 4A). In a classical view, cells in the distal mesenchyme remain in an undifferentiated and proliferative status. Upon leaving the AER's influence, these cells then undergo differentiation and contribute to the different segments of the limb bud. It was considered that the time in which mesenchymal cells remained under the influence of the AER-FGFs was translated into a proximal to distal fate. Hence, a shorter time of influence is translated into more proximal fates while a longer time of the influence results in more distal fates. This simple and classical view was referred to as the "progress zone" model [98], and was accepted by the scientific community for decades.

In recent years, however, genetic evidence involving *Fgf* genes and retinoic acid activities have challenged the progress zone model. Analysis of double knockouts of *Fgf4* and *Fgf8* in the limb ectoderm as well as analysis of cell survival and fate mapping in chick limb buds, have lead to a model in which AER-FGFs promote expansion of mesenchymal cells that have been specified along the proximal-distal axis at early stages of limb development, rather than progressive specification [83, 99]. In this model, termed as the "early specification model", FGF signaling from the AER expands pre-existing proximal-distal territories by controlling survival and promoting proliferation in the distal mesenchymal cells. More recently, by eliminating different combinations of AER-FGFs, it was demonstrated that the AER-FGFs could serve as an instructive signal for proximal-distal patterning [84]. Further thoughts have arisen through re-evaluating the progress zone and early specification models, leading to the hypothesis of the "two-signal gradient model" [100]. This model involves both proximal and distal signals for patterning (Figure 4A). Retinoic acid signaling acts as a proximal signal, which activates *Meis* homeobox transcription factor genes and induce proximal fates in the limb mesenchyme, while the opposing activities of AER-

FGFs serve as distal signals [101-103]. Data obtained from *in vivo* manipulation of chick limb buds and in culture experiments support the hypothesis that proximal mesenchymal cells acquire proximal fates due to the presence of endogenous retinoic acid, and that the balance between distal FGF from the AER and proximal retinoic acid instructs proximal-distal patterning [104, 105]. Although these experiments in chick embryos suggest a role of retinoic acid in proximal-distal patterning, the role of endogenous retinoic acid signaling remains controversial. This is due to the fact that genetic analyses performed in mouse embryos support the hypothesis that endogenous retinoic acid is unlikely to have a role in proximal-distal patterning of the limb bud [106, 107]. Mouse embryos that lack detectable levels of retinoic acid in the limb bud were generated by inactivating the aldehyde dehydrogenase 1 family, member A2 gene (*Aldh1a2*, also called retinaldehyde dehydrogenase 2, *Raldh2*) or a natural mutation of retinol dehydrogenase 10 (*Rdh10*). *Aldh1a2* encodes the rate-limiting enzyme for retinoic acid synthesis and is the most predominant enzyme expressed in developing embryos during the early limb development stages. *Rdh10* encodes an enzyme that converts retinol to retinal, a precursor of retinoic acid. Although embryos lacking retinoic acid arrest in their development before the formation of limb buds [108], developmental arrest can be rescued by dietary administration of retinoic acid to the pregnant mother. In such experiments, levels of retinoic acid in developing embryos were evaluated by the RERE-LacZ transgene, which serves as a reporter of retinoic acid signaling [109] with detection sensitivity of 25 nM retinoic acid [107]. In the absence of detectable levels of retinoic acid with this reporter transgene in the limb bud, proximal-distal patterning seemed to be unaffected. Several potential factors might be involved in such discrepancies, and need to be studied in the future. These include species diversity, differences in the experimental systems (loss or gain of function), and a possibility of the presence and function of undetectable levels of retinoic acid in mutant embryos. In summary, different models have been proposed to explain how AER signals control proximal-distal outgrowth and patterning. Nonetheless, a consensus exists that, regardless of what model is acting, FGF signaling emanating from the AER is a critical regulator in proximal-distal patterning.

Role of AER in Anterior-Posterior Patterning

As described in other reviews [2, 4, 110], anterior-posterior patterning is mediated by *Shh*, which is expressed in the posterior margin of the mesenchyme [74]. The AER participates in the anterior-posterior patterning through the SHH-GREMLIN1-FGF feedback loop, which is critical for maintenance of the AER as described above, as well as maintenance of *Shh* expression. The mechanism by which *Shh* expression is restricted and maintained only in the posterior mesenchyme has been extensively studied and involves SHH activity itself (reviewed in [111, 112]). Recent studies of members of the ETS (E-twenty six) transcription factor family, which are targets of FGF signaling [64, 113], have demonstrated that AER activity is essential for posterior mesenchyme-specific *Shh* expression, and hence, for proper anterior-posterior patterning. Limb mesenchyme-specific conditional inactivation of both *Etv4* and *Etv5*, which are members of the ETS transcription factors and are expressed in the distal mesenchyme of the limb bud, leads to ectopic *Shh* expression in the anterior mesenchyme [114, 115]. Therefore, AER-FGF-dependent activation of *Etv4* and *Etv5* in the mesenchyme prevents the expansion of *Shh* expression in the anterior mesenchyme, allowing for proper anterior-posterior patterning by posteriorly-expressed *Shh* (Figure 4C). This finding not only provides genetic evidence for AER participation in anterior-posterior patterning, but also its mechanism through downstream targets (i.e., *Etv* genes).

AER Activity and Chondrogenic and Apoptotic Signals

During limb development, cells from the limb mesenchyme undergo chondrogenesis or programmed cell death (PCD) (Figure 5), while other cell types, such as muscle, nerves and blood vessels, are derived from trunk structures. The process by which mesenchymal cells are driven towards chondrogenesis or apoptosis is genetically controlled and influenced by AER-mesenchyme interactions [9, 116, 117]. Different PCD areas are responsible for shaping the limbs, however, of particular interest are the interdigital necrotic zones (INZs) of undifferentiated mesenchyme and AER cells that undergo apoptosis [118] (Figure 5). PCD in AER cells is a mechanism to control cell number in the AER and avoid excess AER activity. Analyses to date clearly indicate that changes in PCD are associated with the alteration of gene expression in the AER. Notch signaling is one of the pathways involved in regulating AER activity. Conditional *Notch1* mutant mice display webbed digits as a result of AER hyperplasia and enhanced *Fgf8* expression in the AER [119]. Other molecules involved in the control of PCD are BMPs, which have been studied extensively in relation to PCD.

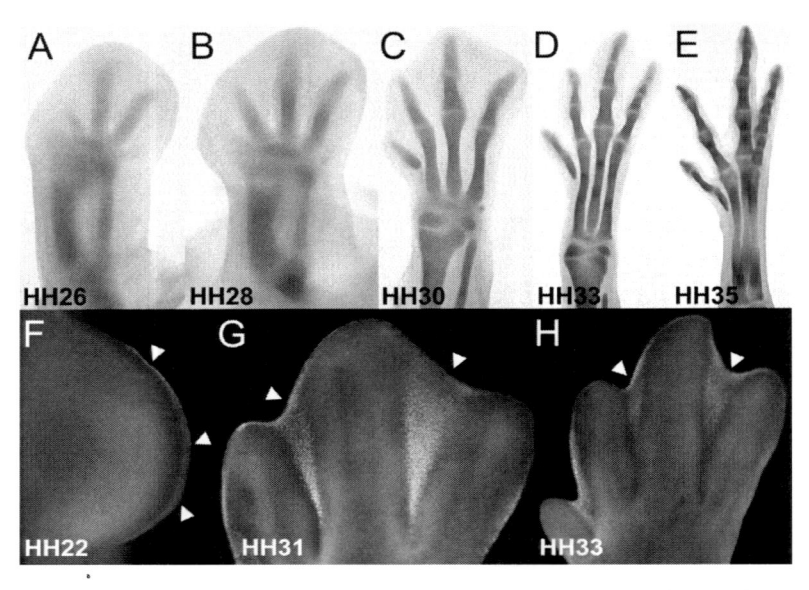

Figure 5. Chondrogenesis and programmed cell death in developing limb buds. (A-E): Chicken hindlimbs stained with Alcian Green to reveal chondrogenic elements. Note the sequencial formation of phalanxes as development proceeds. Limbs are staged according to Hamburguer and Hamilton's (HH) series (162). From A to E, stage HH26, HH28, HH30, HH33 and HH35 limb buds corresponding to 4.5, 5.5, 6.5, 7.5 and 8.5 days of embryonic development. (F-H): Hindlimb buds labeled by Acridine Orange staining to visualize apoptotic cells. (F) Cell death is evident in AER in the stage HH22 hindlimb bud (white arrowheads). Massive cell death in the interdigital areas can be observed at stage HH31 (G) and HH33 (H) hindlimb buds (white arrowheads).

Until recently, the specific role of the AER in PCD was unclear due to the expression of *Bmp2, Bmp4* and *Bmp7* in both the AER and mesenchyme [32]. Genetic experiments in mice, in which gene functions in the AER or AER precursor cells were manipulated, addressed specific functions of genes expressed in the AER during PCD.

As explained in the AER maturation section, early inactivation of *Bmpr1a* in the ectoderm of the limb bud-forming region prevents AER formation [26]. However, when *Bmpr1a* deletion was performed after limb initiation, the AER was formed with an enlarged *Fgf8*-expressing domain with upregulated ectodermal *Fgf4* and *Fgf8* expression. In such limbs, a loss of interdigital PCD and webbed digits were observed [26]. Given the functional redundancy of BMP ligands, inactivation of specific *Bmp* ligand genes, as well as in combination, has recently been performed. AER-specific double mutants for

Bmp2 and *Bmp4* exhibited extra digits, digit bifurcations and webbed limbs due to augmented *Fgf* expression in the AER [35]. Furthermore, triple conditional deletion of *Bmp2*, *Bmp4* and *Bmp7* from the AER resulted in digit patterning defects, including polydactyly, interdigital webbing and splitting digits. These reports demonstrated that BMPs from the AER are not needed for limb outgrowth, but are required only to maintain the AER for proper digit morphogenesis [120]. This interpretation is also in agreement with another recent report, in which Smad1 and Smad5, intracellular effectors of BMP signaling, were inactivated in the AER and the ventral ectoderm. Such genetic manipulation caused syndactyly due to prolonged *Fgf8* expression in the interdigital ectoderm [121]. In conclusion, BMP activity from the AER is essential for proper ectodermal development, and controls PCD in the INZs through regulation of FGF signaling levels emanating from the AER (Figure 4D).

The chondrogenic aggregates that are formed in the limb bud are also controlled by activation of different members of the TGFß superfamily in the mesenchyme. Activins, TGFßs and BMPs are key signaling molecules that have been extensively studied in this process [117]. Phosphorylated Samd1/5/8 is detected as a crescent-like domain at the tip of the digit-forming region, indicating active BMP signaling in this domain [122, 123]. Furthermore, analysis of the phosphorylation of Smad2, a transducer of Activin /TGFß signaling, as well as manipulation of Activin/TGFß signaling through bead implantation, showed a crosstalk between BMP and Activin /TGFß signaling. Since TGFß2 and TGFß3 are produced in the AER [124], these molecules seem to also be involved in modulating BMP function in the AER to regulate digit chondrogenesis (Figure 4D, Figure 5). The presence of latent TGFß Binding protein 1 (LTBP1), which is a key extracellular modulator of TGFß ligand bioavailability, in the AER also suggests complex regulation of BMP activity by TGFß in the AER [124]. Thus, the AER controls digit chondrogenesis and morphogenesis through complex regulation of BMP activities.

THE DEVELOPMENT OF THE LIMB AS AN USEFUL MODEL SYSTEM

As an experimental platform, limb development provides opportunities to address biological questions. In this section, we highlight several characteristic features of general limb development in order to reinforce its usefulness as an experimental system. We then describe several recent studies on mammalian evolutionary diversity, human limb malformation and regeneration, as examples of limb development topics that contribute to diverse biological questions.

Limb Development as a Research Model System

The study of limb development has been a paradigm in the field of developmental biology, making the limb bud a representative and useful organ, widely used to understand a large, diverse collection of processes. Among others, these include cellular and tissue interactions, patterning, organ formation and evolution. The cited works in this review are illustrative examples of limb bud development as a versatile model organ.

Among the processes that can be extensively studied during limb bud development, it is important to highlight the large amount of knowledge generated regarding organ positioning and initiation, growth, patterning, ectodermal/mesenchymal interactions, cell differentiation, apoptosis, chondrogenesis and osteogenesis. Furthermore, the use of limb development as a research model system has also contributed in the understanding of the migration of muscle precursor cells, their differentiation and subsequent muscle formation, nerve growth and guidance, angiogenesis and the contribution of extracellular matrix to development. The combination of chick and mouse limb development, in particular, provides opportunities to perform embryological manipulation, explant culture and powerful genetic approaches. Moreover, limb development studies have helped clarify recently developed mechanisms and/or concepts, such as micro RNA activity during development [125]. In addition, the ease with which limb buds can be dissected offers an opportunity to perform genome-wide, large-scale analyses, including microarrays and chromatin immunoprecipitation (ChIP) assays [126-129]. Furthermore, theoretical and systems biology researchers have used limb development to model developmental processes that can be re-evaluated by experimental approaches [13, 130-134]. Below, by highlighting recent reports in three examples (evolutionary diversity, studies of human disorders and regeneration), we describe how studies of limb development remain attractive and useful in enhancing our knowledge of biological and biomedical sciences.

Evolutionary Aspects of the Vertebrate Limb: Conservation and Diversity

Vertebrate limbs exhibit a large degree of homology in terms of basic structure and underlying developmental systems, indicating a shared evolutionary history of limb development [135]. Structural homology is often seen in the comparison of limbs between species such as the human, bat, bird, and whale.

The basic structural units of the vertebrate limb are conserved: the stylopod (humerus/femur), the zeugopod (radius and ulna/tibia and fibula), and the autopod (carpal-metacarpal/tarsal-metatarsal and digits). Despite this basic structural homology, a large degree of diversity exists in the morphology and growth rates of vertebrate limbs, suggesting differences in the spatial and temporal expression of limb developmental genes. Studies in "non-traditional" mammalian models, such as the bat (*Myotis lucifugus*), opossum (*Monodelphis domestica*), and tammar wallaby (*Macropus eugenii*) have begun to reveal differences in limb gene expression and developmental processes [136].

The elongated digits and limbs possessed by the bat have been linked to both an expanded AER and *Fgf8* expression domain [137, 138], while, despite conserved *Fgf8* expression, no physical, mature AER was found in the opossum, which displays highly developed forelimbs at birth [139, 140].

In contrast, marsupials, such as the wallaby, possess hindlimbs that are much larger than the forelimb and contain syndactylous digits. Recent insight into marsupial limbs has revealed that the *HoxA13* and *HoxD13* genes are important for the differences in limb size and digit patterning [141]. Establishing non-traditional model animals for developmental studies will help in understanding the common and species-specific morphological, genetic and molecular systems required for limb development, and further allow us to understand how species diversity is generated throughout vertebrate evolution [142].

Human Limb Malformations

Continued research into limb developmental processes will help explain the diverse morphologies seen within human congenital limb formations and the genetic causes behind them. Congenital limb malformations are found in one in 200-500 live births [143], and patients with limb malformations often exhibit defects in other organs, such as the heart, kidney, skin and central nervous system. Recently, research in limb development has contributed to advancements in the knowledge of various human limb abnormalities.

In the case of the human forelimb, recent studies have linked cases of brachydactyly and polydactyly of the forelimb digits to aberrant *Wnt* and *Shh* signaling, respectively [144, 145]. Hindlimb congenital malformations include fusion of the legs (sirenomelia), clubfoot, and polydactyly. Mice displaying sirenomelia were found to be lacking the enzyme Cyp26a1, resulting in excess retinoic acid signaling and deficient BMP signaling in the body [146]. Mutations in *PITX1*, a transcription factor expressed primarily in the developing limb, were also found to be associated with individuals displaying clubfoot and mirror-image polydactyly [147-149]. Understanding limb development, as well as unveiling the genetic, molecular and cellular mechanisms that regulate limb development, can contribute to not only enhancing our knowledge of limb development but also understanding the mechanisms behind congenital defects of other organs.

AER and Regeneration

Animal models, including zebrafish, axolotl and *Xenopus*, have been widely used for the study of appendage regeneration, such as the limb and fin [150]. Contrary to studies in adult animals, the use of developing chick and mouse embryos indicates that embryonic limbs are also a useful system to study regeneration. During limb/fin regeneration, epidermal wound healing culminates with the formation of an apical epithelial cap (AEC), an essential tissue for regeneration. The AEC is a specialized, multilayered epidermal structure, which corresponds to the AER of developing limb buds [150-153]. Chick limb buds do not regenerate after experimental amputation [80-82]. However, activating Wnt/ß-catenin signaling and FGF signaling can induce regeneration of the limb bud and is associated with restoration of the AER of the limb bud [153, 154]. Similarly, activating these signaling pathways can enhance the ability of limb regeneration in larvae and adult animals, which involves activating the epidermal responses [154, 155]. In the case of the mouse, embryos at E14.5 are able to regenerate the last phalanx after distal amputation [156-158]. Similar to limb development, BMP4 and the transcription factor *Msx1* are involved in digit tip regeneration [157, 159-161].

Although developmental processes in embryonic limbs are not completely identical to those in limb regeneration in adult animals, embryonic limb development and, in particular, studies of induction, maturation, maintenance and regression of the AER would offer a simple and amenable experimental platform, which could be applied to the study of limb regeneration.

CONCLUSION

The AER is a specialized ectodermal structure, transiently present in developing limb buds. Recent studies have contributed to the understanding of the molecular and genetic mechanisms of its emergence, maturation, maintenance and regression.

The role of the AER during limb development includes patterning along the proximal-distal (shoulder to finger) axis and anterior-posterior (thumb to little finger) axis. In addition, the AER functions in shaping the developing limb through regulating chondrogenesis and apoptosis. The knowledge gained from our understanding of the AER and its roles during vertebrate limb development has the potential to explain a wide variety of unanswered questions, ranging from species diversity to human malformations, stressing the need for continued research into the AER, and in general, the limb itself, as we move into the future.

ACKNOWLEDGMENTS

Work in the laboratory of YK is supported in part by the Minnesota Medical Foundation (3962-9211-09) and the National Institute of Arthritis and Musculoskeletal and Skin Diseases of the National Institutes of Health (AR063782-01). Work in the laboratory of JRL is supported by a grant from Fundação para a Ciência e a Tecnologia (FCT), Portugal (PTDC/BIA-BCM/100867/2008). JRL is supported by the Subprogram Ramón y Cajal from Ministerio de Ciencia e Innovación, Spain (RYC-2008-02753).

REFERENCES

[1] Duboc V, Logan MP. Regulation of limb bud initiation and limb-type morphology. *Dev. Dyn.,* 2011 May;240(5):1017-27.

[2] McGlinn E, Tabin CJ. Mechanistic insight into how Shh patterns the vertebrate limb. *Current opinion in genetics and development,* 2006 Aug;16(4):426-32.
[3] Niswander L. Pattern formation: old models out on a limb. *Nat. Rev. Genet.,* 2003 Feb;4(2):133-43.
[4] Zeller R, Lopez-Rios J, Zuniga A. Vertebrate limb bud development: moving towards integrative analysis of organogenesis. *Nat. Rev. Genet.,* 2009 Dec;10(12):845-58.
[5] Towers M, Tickle C. Growing models of vertebrate limb development. Development (Cambridge, England). 2009 Jan;136(2):179-90.
[6] Fernandez-Teran M, Ros MA. The Apical Ectodermal Ridge: morphological aspects and signaling pathways. *The International journal of developmental biology,* 2008;52(7):857-71.
[7] Zakany J, Duboule D. The role of Hox genes during vertebrate limb development. *Current opinion in genetics and development,* 2007 Aug;17(4):359-66.
[8] Hopyan S, Sharpe J, Yang Y. Budding behaviors: Growth of the limb as a model of morphogenesis. *Dev. Dyn.,* 2011 May;240(5):1054-62.
[9] Montero JA, Hurle JM. Sculpturing digit shape by cell death. *Apoptosis,* 2010 Mar;15(3):365-75.
[10] Shubin N, Tabin C, Carroll S. Deep homology and the origins of evolutionary novelty. *Nature,* 2009 Feb 12;457(7231):818-23.
[11] Don EK, Currie PD, Cole NJ. The evolutionary history of the development of the pelvic fin/hindlimb. *Journal of anatomy,* 2012 Aug 23.
[12] Searls RL, Janners MY. The initiation of limb bud outgrowth in the embryonic chick. *Developmental biology,* 1971 Feb;24(2):198-213.
[13] Damon BJ, Mezentseva NV, Kumaratilake JS, Forgacs G, Newman SA. Limb bud and flank mesoderm have distinct "physical phenotypes" that may contribute to limb budding. *Developmental biology,* 2008 Sep 15;321(2):319-30.
[14] Wyngaarden LA, Vogeli KM, Ciruna BG, Wells M, Hadjantonakis AK, Hopyan S. Oriented cell motility and division underlie early limb bud morphogenesis. *Development,* (Cambridge, England). 2010 Aug 1;137(15):2551-8.
[15] Gros J, Hu JK, Vinegoni C, Feruglio PF, Weissleder R, Tabin CJ. WNT5A/JNK and FGF/MAPK pathways regulate the cellular events shaping the vertebrate limb bud. *Curr. Biol.,* 2010 Nov 23;20(22):1993-2002.

[16] Altabef M, Clarke JD, Tickle C. Dorso-ventral ectodermal compartments and origin of apical ectodermal ridge in developing chick limb. *Development,* (Cambridge, England). 1997 Nov;124(22):4547-56.
[17] Kimmel RA, Turnbull DH, Blanquet V, Wurst W, Loomis CA, Joyner AL. Two lineage boundaries coordinate vertebrate apical ectodermal ridge formation. *Genes and development,* 2000 Jun 1;14(11):1377-89.
[18] Sekine K, Ohuchi H, Fujiwara M, Yamasaki M, Yoshizawa T, Sato T, et al. Fgf10 is essential for limb and lung formation. *Nature genetics,* 1999 Jan;21(1):138-41.
[19] Min H, Danilenko DM, Scully SA, Bolon B, Ring BD, Tarpley JE, et al. Fgf-10 is required for both limb and lung development and exhibits striking functional similarity to Drosophila branchless. *Genes and development,* 1998 Oct 15;12(20):3156-61.
[20] Kawakami Y, Capdevila J, Buscher D, Itoh T, Rodriguez Esteban C, Izpisua Belmonte JC. WNT signals control FGF-dependent limb initiation and AER induction in the chick embryo. *Cell,* 2001 Mar 23;104(6):891-900.
[21] Kengaku M, Capdevila J, Rodriguez-Esteban C, De La Pena J, Johnson RL, Belmonte JC, et al. Distinct WNT pathways regulating AER formation and dorsoventral polarity in the chick limb bud. *Science,* (New York, NY. 1998;280(5367):1274-7.
[22] Barrow JR, Thomas KR, Boussadia-Zahui O, Moore R, Kemler R, Capecchi MR, et al. Ectodermal Wnt3/beta-catenin signaling is required for the establishment and maintenance of the apical ectodermal ridge. *Genes and development,* 2003 Feb 1;17(3):394-409.
[23] Soshnikova N, Zechner D, Huelsken J, Mishina Y, Behringer RR, Taketo MM, et al. Genetic interaction between Wnt/beta-catenin and BMP receptor signaling during formation of the AER and the dorsal-ventral axis in the limb. *Genes and development,* 2003 Aug 15;17(16):1963-8.
[24] Arman E, Haffner-Krausz R, Gorivodsky M, Lonai P. Fgfr2 is required for limb outgrowth and lung-branching morphogenesis. Proceedings of the National Academy of Sciences of the United States of America. 1999 Oct 12;96(21):11895-9.
[25] Xu X, Weinstein M, Li C, Naski M, Cohen RI, Ornitz DM, et al. Fibroblast growth factor receptor 2 (FGFR2)-mediated reciprocal regulation loop between FGF8 and FGF10 is essential for limb induction. *Development,* (Cambridge, England). 1998 Feb;125(4):753-65.

[26] Pajni-Underwood S, Wilson CP, Elder C, Mishina Y, Lewandoski M. BMP signals control limb bud interdigital programmed cell death by regulating FGF signaling. *Development,* (Cambridge, England). 2007 Jun;134(12):2359-68.

[27] Ahn K, Mishina Y, Hanks MC, Behringer RR, Crenshaw EB, 3rd. BMPR-IA signaling is required for the formation of the apical ectodermal ridge and dorsal-ventral patterning of the limb. *Development,* (Cambridge, England). 2001 Nov;128(22):4449-61.

[28] Joyner AL, Martin GR. En-1 and En-2, two mouse genes with sequence homology to the Drosophila engrailed gene: expression during embryogenesis. *Genes and development,* 1987 Mar;1(1):29-38.

[29] Loomis CA, Kimmel RA, Tong CX, Michaud J, Joyner AL. Analysis of the genetic pathway leading to formation of ectopic apical ectodermal ridges in mouse Engrailed-1 mutant limbs. *Development,* (Cambridge, England). 1998 Mar;125(6):1137-48.

[30] Logan C, Hornbruch A, Campbell I, Lumsden A. The role of Engrailed in establishing the dorsoventral axis of the chick limb. *Development,* (Cambridge, England). 1997 Jun;124(12):2317-24.

[31] Pizette S, Abate-Shen C, Niswander L. BMP controls proximodistal outgrowth, via induction of the apical ectodermal ridge, and dorsoventral patterning in the vertebrate limb. *Development,* (Cambridge, England). 2001 Nov;128(22):4463-74.

[32] Robert B. Bone morphogenetic protein signaling in limb outgrowth and patterning. Development, *growth and differentiation,* 2007 Aug;49(6):455-68.

[33] Wang CK, Omi M, Ferrari D, Cheng HC, Lizarraga G, Chin HJ, et al. Function of BMPs in the apical ectoderm of the developing mouse limb. *Developmental biology,* 2004 May 1;269(1):109-22.

[34] Plikus M, Wang WP, Liu J, Wang X, Jiang TX, Chuong CM. Morphoregulation of ectodermal organs: integument pathology and phenotypic variations in K14-Noggin engineered mice through modulation of bone morphogenic protein pathway. *The American journal of pathology,* 2004 Mar;164(3):1099-114.

[35] Maatouk DM, Choi KS, Bouldin CM, Harfe BD. In the limb AER Bmp2 and Bmp4 are required for dorsal-ventral patterning and interdigital cell death but not limb outgrowth. *Developmental biology,* 2009 Mar 15;327(2):516-23.

[36] Yang A, Schweitzer R, Sun D, Kaghad M, Walker N, Bronson RT, et al. p63 is essential for regenerative proliferation in limb, craniofacial and epithelial development. *Nature,* 1999 Apr 22;398(6729):714-8.

[37] Mills AA, Zheng B, Wang XJ, Vogel H, Roop DR, Bradley A. p63 is a p53 homologue required for limb and epidermal morphogenesis. *Nature,* 1999 Apr 22;398(6729):708-13.

[38] Bell SM, Schreiner CM, Waclaw RR, Campbell K, Potter SS, Scott WJ. Sp8 is crucial for limb outgrowth and neuropore closure. Proceedings of the National Academy of Sciences of the United States of America. 2003 Oct 2.

[39] Treichel D, Schock F, Jackle H, Gruss P, Mansouri A. mBtd is required to maintain signaling during murine limb development. *Genes and development,* 2003 Nov 1;17(21):2630-5.

[40] Casanova JC, Uribe V, Badia-Careaga C, Giovinazzo G, Torres M, Sanz-Ezquerro JJ. Apical ectodermal ridge morphogenesis in limb development is controlled by Arid3b-mediated regulation of cell movements. *Development,* (Cambridge, England). 2011 Mar;138(6):1195-205.

[41] Kawakami Y, Esteban CR, Matsui T, Rodriguez-Leon J, Kato S, Belmonte JC. Sp8 and Sp9, two closely related buttonhead-like transcription factors, regulate Fgf8 expression and limb outgrowth in vertebrate embryos. *Development,* (Cambridge, England). 2004 Oct;131(19):4763-74.

[42] Talamillo A, Delgado I, Nakamura T, de-Vega S, Yoshitomi Y, Unda F, et al. Role of Epiprofin, a zinc-finger transcription factor, in limb development. *Developmental biology,* 2010 Jan 15;337(2):363-74.

[43] Barbieri CE, Tang LJ, Brown KA, Pietenpol JA. Loss of p63 leads to increased cell migration and up-regulation of genes involved in invasion and metastasis. *Cancer Res.,* 2006 Aug 1;66(15):7589-97.

[44] Carroll DK, Carroll JS, Leong CO, Cheng F, Brown M, Mills AA, et al. p63 regulates an adhesion programme and cell survival in epithelial cells. *Nature cell biology,* 2006 Jun;8(6):551-61.

[45] Testoni B, Borrelli S, Tenedini E, Alotto D, Castagnoli C, Piccolo S, et al. Identification of new p63 targets in human keratinocytes. *Cell Cycle,* 2006 Dec;5(23):2805-11.

[46] Vigano MA, Lamartine J, Testoni B, Merico D, Alotto D, Castagnoli C, et al. New p63 targets in keratinocytes identified by a genome-wide approach. *The EMBO journal,* 2006 Nov 1;25(21):5105-16.

[47] Shimomura Y, Wajid M, Shapiro L, Christiano AM. P-cadherin is a p63 target gene with a crucial role in the developing human limb bud and hair follicle. *Development,* (Cambridge, England). 2008 Feb;135(4):743-53.

[48] Laufer E, Nelson CE, Johnson RL, Morgan BA, Tabin C. Sonic hedgehog and Fgf-4 act through a signaling cascade and feedback loop to integrate growth and patterning of the developing limb bud. *Cell,* 1994 Dec 16;79(6):993-1003.

[49] Zuniga A, Haramis AP, McMahon AP, Zeller R. Signal relay by BMP antagonism controls the SHH/FGF4 feedback loop in vertebrate limb buds. *Nature,* 1999 Oct 7;401(6753):598-602.

[50] Khokha MK, Hsu D, Brunet LJ, Dionne MS, Harland RM. Gremlin is the BMP antagonist required for maintenance of Shh and Fgf signals during limb patterning. *Nature genetics,* 2003 Jul;34(3):303-7.

[51] Panman L, Galli A, Lagarde N, Michos O, Soete G, Zuniga A, et al. Differential regulation of gene expression in the digit forming area of the mouse limb bud by SHH and gremlin 1/FGF-mediated epithelial-mesenchymal signalling. *Development,* (Cambridge, England). 2006 Sep;133(17):3419-28.

[52] Benazet JD, Bischofberger M, Tiecke E, Goncalves A, Martin JF, Zuniga A, et al. A self-regulatory system of interlinked signaling feedback loops controls mouse limb patterning. *Science,* (New York, NY. 2009 Feb 20;323(5917):1050-3.

[53] Michos O, Panman L, Vintersten K, Beier K, Zeller R, Zuniga A. Gremlin-mediated BMP antagonism induces the epithelial-mesenchymal feedback signaling controlling metanephric kidney and limb organogenesis. *Development,* (Cambridge, England). 2004 Jul;131(14):3401-10.

[54] Ohuchi H, Nakagawa T, Yamamoto A, Araga A, Ohata T, Ishimaru Y, et al. The mesenchymal factor, FGF10, initiates and maintains the outgrowth of the chick limb bud through interaction with FGF8, an apical ectodermal factor. *Development,* (Cambridge, England). 1997 Jun;124(11):2235-44.

[55] Jho EH, Zhang T, Domon C, Joo CK, Freund JN, Costantini F. Wnt/beta-catenin/Tcf signaling induces the transcription of Axin2, a negative regulator of the signaling pathway. *Molecular and cellular biology,* 2002 Feb;22(4):1172-83.

[56] DasGupta R, Fuchs E. Multiple roles for activated LEF/TCF transcription complexes during hair follicle development and

differentiation. *Development,* (Cambridge, England). 1999 Oct;126(20):4557-68.

[57] Maretto S, Cordenonsi M, Dupont S, Braghetta P, Broccoli V, Hassan AB, et al. Mapping Wnt/beta-catenin signaling during mouse development and in colorectal tumors. Proceedings of the National Academy of Sciences of the United States of America. 2003 Mar 18;100(6):3299-304.

[58] Ahrens MJ, Romereim S, Dudley AT. A re-evaluation of two key reagents for in vivo studies of Wnt signaling. *Dev. Dyn.,* 2011 Sep;240(9):2060-8.

[59] Nam JS, Turcotte TJ, Yoon JK. Dynamic expression of R-spondin family genes in mouse development. *Gene. Expr. Patterns,* 2007 Jan;7(3):306-12.

[60] Bell SM, Schreiner CM, Wert SE, Mucenski ML, Scott WJ, Whitsett JA. R-spondin 2 is required for normal laryngeal-tracheal, lung and limb morphogenesis. *Development,* (Cambridge, England). 2008 Mar;135(6):1049-58.

[61] Nam JS, Park E, Turcotte TJ, Palencia S, Zhan X, Lee J, et al. Mouse R-spondin2 is required for apical ectodermal ridge maintenance in the hindlimb. *Developmental biology,* 2007 Nov 1;311(1):124-35.

[62] Aoki M, Kiyonari H, Nakamura H, Okamoto H. R-spondin2 expression in the apical ectodermal ridge is essential for outgrowth and patterning in mouse limb development. *Development, growth and differentiation,* 2008 Feb;50(2):85-95.

[63] Lu P, Yu Y, Perdue Y, Werb Z. The apical ectodermal ridge is a timer for generating distal limb progenitors. *Development,* (Cambridge, England). 2008 Apr;135(8):1395-405.

[64] Kawakami Y, Rodriguez-Leon J, Koth CM, Buscher D, Itoh T, Raya A, et al. MKP3 mediates the cellular response to FGF8 signalling in the vertebrate limb. *Nature cell biology,* 2003 Jun;5(6):513-9.

[65] Tomas AR, Certal AC, Rodriguez-Leon J. FLRT3 as a key player on chick limb development. *Developmental biology,* 2011 Jul 15;355(2):324-33.

[66] Bottcher RT, Pollet N, Delius H, Niehrs C. The transmembrane protein XFLRT3 forms a complex with FGF receptors and promotes FGF signalling. *Nature cell biology,* 2004 Jan;6(1):38-44.

[67] Haines BP, Wheldon LM, Summerbell D, Heath JK, Rigby PW. Regulated expression of FLRT genes implies a functional role in the

regulation of FGF signalling during mouse development. *Developmental biology,* 2006 Sep 1;297(1):14-25.

[68] Maretto S, Muller PS, Aricescu AR, Cho KW, Bikoff EK, Robertson EJ. Ventral closure, headfold fusion and definitive endoderm migration defects in mouse embryos lacking the fibronectin leucine-rich transmembrane protein FLRT3. *Developmental biology,* 2008 Jun 1;318(1):184-93.

[69] Pearse RV, 2nd, Vogan KJ, Tabin CJ. Ptc1 and Ptc2 transcripts provide distinct readouts of Hedgehog signaling activity during chick embryogenesis. *Developmental biology,* 2001 Nov 1;239(1):15-29.

[70] Quirk J, van den Heuvel M, Henrique D, Marigo V, Jones TA, Tabin C, et al. The smoothened gene and hedgehog signal transduction in Drosophila and vertebrate development. *Cold Spring Harb. Symp. Quant. Biol.,* 1997;62:217-26.

[71] Bell SM, Schreiner CM, Scott WJ. Disrupting the establishment of polarizing activity by teratogen exposure. *Mechanisms of development,* 1999 Nov;88(2):147-57.

[72] Bell SM, Schreiner CM, Goetz JA, Robbins DJ, Scott WJ, Jr. Shh signaling in limb bud ectoderm: potential role in teratogen-induced postaxial ectrodactyly. *Dev. Dyn.,* 2005 Jun;233(2):313-25.

[73] Bouldin CM, Gritli-Linde A, Ahn S, Harfe BD. Shh pathway activation is present and required within the vertebrate limb bud apical ectodermal ridge for normal autopod patterning. Proceedings of the National Academy of Sciences of the United States of America. 2010 Mar 23;107(12):5489-94.

[74] Riddle RD, Johnson RL, Laufer E, Tabin C. Sonic hedgehog mediates the polarizing activity of the ZPA. *Cell,* 1993 Dec 31;75(7):1401-16.

[75] Hirashima T, Iwasa Y, Morishita Y. Distance between AER and ZPA is defined by feed-forward loop and is stabilized by their feedback loop in vertebrate limb bud. *Bull. Math. Biol.,* 2008 Feb;70(2):438-59.

[76] Tamura K, Nomura N, Seki R, Yonei-Tamura S, Yokoyama H. Embryological evidence identifies wing digits in birds as digits 1, 2, and 3. *Science,* (New York, NY. 2011 Feb 11;331(6018):753-7.

[77] Wang CK, Tsugane MH, Scranton V, Kosher RA, Pierro LJ, Upholt WB, et al. Pleiotropic patterning response to activation of Shh signaling in the limb apical ectodermal ridge. *Dev. Dyn.,* 2011 May;240(5):1289-302.

[78] Scherz PJ, Harfe BD, McMahon AP, Tabin CJ. The limb bud Shh-Fgf feedback loop is terminated by expansion of former ZPA cells. *Science*, (New York, NY. 2004 Jul 16;305(5682):396-9.

[79] Verheyden JM, Sun X. An Fgf/Gremlin inhibitory feedback loop triggers termination of limb bud outgrowth. *Nature*, 2008 Jun 25.

[80] Saunders JW, Jr. The proximo-distal sequence of origin of the parts of the chick wing and the role of the ectoderm. *J. Exp. Zool.*, 1948 Aug;108(3):363-403.

[81] Summerbell D. A quantitative analysis of the effect of excision of the AER from the chick limb-bud. *J. Embryol. Exp. Morphol.*, 1974 Dec;32(3):651-60.

[82] Niswander L, Tickle C, Vogel A, Booth I, Martin GR. FGF-4 replaces the apical ectodermal ridge and directs outgrowth and patterning of the limb. *Cell*, 1993;75(3):579-87.

[83] Sun X, Mariani FV, Martin GR. Functions of FGF signalling from the apical ectodermal ridge in limb development. *Nature*, 2002 Aug 1;418(6897):501-8.

[84] Mariani FV, Ahn CP, Martin GR. Genetic evidence that FGFs have an instructive role in limb proximal-distal patterning. *Nature*, 2008 May 15;453(7193):401-5.

[85] Kurose H, Bito T, Adachi T, Shimizu M, Noji S, Ohuchi H. Expression of Fibroblast growth factor 19 (Fgf19) during chicken embryogenesis and eye development, compared with Fgf15 expression in the mouse. *Gene. Expr. Patterns*, 2004 Oct;4(6):687-93.

[86] Wright TJ, Ladher R, McWhirter J, Murre C, Schoenwolf GC, Mansour SL. Mouse FGF15 is the ortholog of human and chick FGF19, but is not uniquely required for otic induction. *Developmental biology*, 2004 May 1;269(1):264-75.

[87] Crossley PH, Martin GR. The mouse Fgf8 gene encodes a family of polypeptides and is expressed in regions that direct outgrowth and patterning in the developing embryo. *Development*, (Cambridge, England). 1995;121(2):439-51.

[88] Ohuchi H, Yoshioka H, Tanaka A, Kawakami Y, Nohno T, Noji S. Involvement of androgen-induced growth factor (FGF-8) gene in mouse embryogenesis and morphogenesis. *Biochemical and biophysical research communications*, 1994 Oct 28;204(2):882-8.

[89] Sun X, Lewandoski M, Meyers EN, Liu YH, Maxson RE, Jr., Martin GR. Conditional inactivation of Fgf4 reveals complexity of signalling during limb bud development. *Nature genetics*, 2000 May;25(1):83-6.

[90] Moon AM, Boulet AM, Capecchi MR. Normal limb development in conditional mutants of Fgf4. *Development,* (Cambridge, England). 2000 Mar;127(5):989-96.
[91] Moon AM, Capecchi MR. Fgf8 is required for outgrowth and patterning of the limbs. *Nature genetics,* 2000 Dec;26(4):455-9.
[92] Lewandoski M, Sun X, Martin GR. Fgf8 signalling from the AER is essential for normal limb development. *Nature genetics,* 2000;26(4):460-3.
[93] Colvin JS, Feldman B, Nadeau JH, Goldfarb M, Ornitz DM. Genomic organization and embryonic expression of the mouse fibroblast growth factor 9 gene. *Dev. Dyn.,* 1999 Sep;216(1):72-88.
[94] Heikinheimo M, Lawshe A, Shackleford GM, Wilson DB, MacArthur CA. Fgf-8 expression in the post-gastrulation mouse suggests roles in the development of the face, limbs and central nervous system. *Mechanisms of development,* 1994 Nov;48(2):129-38.
[95] Mahmood R, Bresnick J, Hornbruch A, Mahony C, Morton N, Colquhoun K, et al. A role for FGF-8 in the initiation and maintenance of vertebrate limb bud outgrowth. *Curr. Biol.,* 1995 Jul 1;5(7):797-806.
[96] Niswander L, Martin GR. Fgf-4 expression during gastrulation, myogenesis, limb and tooth development in the mouse. *Development,* (Cambridge, England). 1992 Mar;114(3):755-68.
[97] Sun X, Mariani FV, Martin GR. Functions of FGF signalling from the apical ectodermal ridge in limb development. *Nature,* 2002;418 (6897):501-8.
[98] Summerbell D, Lewis JH, Wolpert L. Positional information in chick limb morphogenesis. *Nature,* 1973 Aug 24;244(5417):492-6.
[99] Dudley AT, Ros MA, Tabin CJ. A re-examination of proximodistal patterning during vertebrate limb development. *Nature,* 2002;418 (6897):539-44.
[100] Tabin C, Wolpert L. Rethinking the proximodistal axis of the vertebrate limb in the molecular era. *Genes and development,* 2007 Jun 15;21(12):1433-42.
[101] Mercader N, Leonardo E, Azpiazu N, Serrano A, Morata G, Martinez C, et al. Conserved regulation of proximodistal limb axis development by Meis1/Hth. *Nature,* 1999 Nov 25;402(6760):425-9.
[102] Mercader N, Leonardo E, Piedra ME, Martinez AC, Ros MA, Torres M. Opposing RA and FGF signals control proximodistal vertebrate limb development through regulation of Meis genes. *Development,* (Cambridge, England). 2000 Sep;127(18):3961-70.

[103] Yashiro K, Zhao X, Uehara M, Yamashita K, Nishijima M, Nishino J, et al. Regulation of retinoic acid distribution is required for proximodistal patterning and outgrowth of the developing mouse limb. *Developmental cell,* 2004 Mar;6(3):411-22.
[104] Cooper KL, Hu JK, ten Berge D, Fernandez-Teran M, Ros MA, Tabin CJ. Initiation of proximal-distal patterning in the vertebrate limb by signals and growth. *Science,* (New York, NY. 2011 May 27;332 (6033):1083-6.
[105] Rosello-Diez A, Ros MA, Torres M. Diffusible signals, not autonomous mechanisms, determine the main proximodistal limb subdivision. *Science,* (New York, NY. 2011 May 27;332(6033):1086-8.
[106] Zhao X, Sirbu IO, Mic FA, Molotkova N, Molotkov A, Kumar S, et al. Retinoic acid promotes limb induction through effects on body axis extension but is unnecessary for limb patterning. *Curr. Biol.,* 2009 Jun 23;19(12):1050-7.
[107] Cunningham TJ, Chatzi C, Sandell LL, Trainor PA, Duester G. Rdh10 mutants deficient in limb field retinoic acid signaling exhibit normal limb patterning but display interdigital webbing. *Dev. Dyn.,* 2011 May;240(5):1142-50.
[108] Niederreither K, Subbarayan V, Dolle P, Chambon P. Embryonic retinoic acid synthesis is essential for early mouse post-implantation development. *Nature genetics,* 1999 Apr;21(4):444-8.
[109] Rossant J, Zirngibl R, Cado D, Shago M, Giguere V. Expression of a retinoic acid response element-hsplacZ transgene defines specific domains of transcriptional activity during mouse embryogenesis. *Genes and development,* 1991 Aug;5(8):1333-44.
[110] Tickle C. Making digit patterns in the vertebrate limb. *Nature reviews,* 2006 Jan;7(1):45-53.
[111] Benazet JD, Zeller R. Vertebrate limb development: moving from classical morphogen gradients to an integrated 4-dimensional patterning system. *Cold Spring Harb. Perspect. Biol.,* 2009 Oct;1(4):a001339.
[112] Varjosalo M, Taipale J. Hedgehog: functions and mechanisms. *Genes and development,* 2008 Sep 15;22(18):2454-72.
[113] Yordy JS, Muise-Helmericks RC. Signal transduction and the Ets family of transcription factors. *Oncogene,* 2000;19(55):6503-13.
[114] Mao J, McGlinn E, Huang P, Tabin CJ, McMahon AP. Fgf-dependent Etv4/5 activity is required for posterior restriction of Sonic Hedgehog and promoting outgrowth of the vertebrate limb. *Developmental cell,* 2009 Apr;16(4):600-6.

[115] Zhang Z, Verheyden JM, Hassell JA, Sun X. FGF-regulated Etv genes are essential for repressing Shh expression in mouse limb buds. *Developmental cell,* 2009 Apr;16(4):607-13.
[116] Fernandez-Teran MA, Hinchliffe JR, Ros MA. Birth and death of cells in limb development: a mapping study. *Dev. Dyn.,* 2006 Sep;235(9):2521-37.
[117] Montero JA, Hurle JM. Deconstructing digit chondrogenesis. *Bioessays,* 2007 Aug;29(8):725-37.
[118] Zuzarte-Luis V, Hurle JM. Programmed cell death in the embryonic vertebrate limb. *Seminars in cell and developmental biology,* 2005 Apr;16(2):261-9.
[119] Francis JC, Radtke F, Logan MP. Notch1 signals through Jagged2 to regulate apoptosis in the apical ectodermal ridge of the developing limb bud. *Dev. Dyn.,* 2005 Dec;234(4):1006-15.
[120] Choi KS, Lee C, Maatouk DM, Harfe BD. Bmp2, Bmp4 and Bmp7 are co-required in the mouse AER for normal digit patterning but not limb outgrowth. *PloS one,* 2012;7(5):e37826.
[121] Wong YL, Behringer RR, Kwan KM. Smad1/Smad5 signaling in limb ectoderm functions redundantly and is required for interdigital programmed cell death. *Developmental biology,* 2012 Mar 1;363(1):247-57.
[122] Suzuki T, Hasso SM, Fallon JF. Unique SMAD1/5/8 activity at the phalanx-forming region determines digit identity. Proceedings of the National Academy of Sciences of the United States of America. 2008 Mar 18;105(11):4185-90.
[123] Montero JA, Lorda-Diez CI, Ganan Y, Macias D, Hurle JM. Activin/TGFbeta and BMP crosstalk determines digit chondrogenesis. *Developmental biology,.* 2008 Sep 15;321(2):343-56.
[124] Lorda-Diez CI, Montero JA, Garcia-Porrero JA, Hurle JM. Tgfbeta2 and 3 are coexpressed with their extracellular regulator Ltbp1 in the early limb bud and modulate mesodermal outgrowth and BMP signaling in chicken embryos. *BMC developmental biology,* 2010;10:69.
[125] Hornstein E, Mansfield JH, Yekta S, Hu JK, Harfe BD, McManus MT, et al. The microRNA miR-196 acts upstream of Hoxb8 and Shh in limb development. *Nature,* 2005 Dec 1;438(7068):671-4.
[126] Vokes SA, Ji H, Wong WH, McMahon AP. A genome-scale analysis of the cis-regulatory circuitry underlying sonic hedgehog-mediated patterning of the mammalian limb. *Genes and development,* 2008 Oct 1;22(19):2651-63.

[127] Jumlongras D, Lachke SA, O'Connell DJ, Aboukhalil A, Li X, Choe SE, et al. An evolutionarily conserved enhancer regulates Bmp4 expression in developing incisor and limb bud. *PloS one,* 2012;7(6):e38568.

[128] Lettice LA, Williamson I, Wiltshire JH, Peluso S, Devenney PS, Hill AE, et al. Opposing functions of the ETS factor family define Shh spatial expression in limb buds and underlie polydactyly. *Developmental cell,* 2012 Feb 14;22(2):459-67.

[129] Probst S, Kraemer C, Demougin P, Sheth R, Martin GR, Shiratori H, et al. SHH propagates distal limb bud development by enhancing CYP26B1-mediated retinoic acid clearance via AER-FGF signalling. *Development,* (Cambridge, England). 2011 May;138(10):1913-23.

[130] Morishita Y, Iwasa Y. Growth based morphogenesis of vertebrate limb bud. *Bull. Math. Biol.,* 2008 Oct;70(7):1957-78.

[131] Morishita Y, Iwasa Y. Estimating the spatiotemporal pattern of volumetric growth rate from fate maps in chick limb development. *Dev. Dyn.,* 2009 Feb;238(2):415-22.

[132] Glimm T, Zhang J, Shen YQ, Newman SA. Reaction-diffusion systems and external morphogen gradients: the two-dimensional case, with an application to skeletal pattern formation. *Bull. Math. Biol.,* 2012 Mar;74(3):666-87.

[133] Marcon L, Arques CG, Torres MS, Sharpe J. A computational clonal analysis of the developing mouse limb bud. *PLoS Comput. Biol.,* 2011;7(2):e1001071.

[134] Boehm B, Rautschka M, Quintana L, Raspopovic J, Jan Z, Sharpe J. A landmark-free morphometric staging system for the mouse limb bud. *Development,* (Cambridge, England). 2011 Mar;138(6):1227-34.

[135] Abbasi AA. Evolution of vertebrate appendicular structures: Insight from genetic and palaeontological data. *Dev. Dyn.,* 2011 May;240(5):1005-16.

[136] Sears KE. Novel insights into the regulation of limb development from 'natural' mammalian mutants. Studies in 'non-traditional' mammalian models with very different limb morphologies and sizes can contribute to resolving general developmental mechanisms. *Bioessays,* 2011 May;33(5):327-31.

[137] Cretekos CJ, Wang Y, Green ED, Martin JF, Rasweiler JJt, Behringer RR. Regulatory divergence modifies limb length between mammals. *Genes and development,*2008 Jan 15;22(2):141-51.

[138] Hockman D, Mason MK, Jacobs DS, Illing N. The role of early development in mammalian limb diversification: a descriptive

comparison of early limb development between the Natal long-fingered bat (Miniopterus natalensis) and the mouse (Mus musculus). *Dev. Dyn.*, 2009 Apr;238(4):965-79.

[139] Doroba CK, Sears KE. The divergent development of the apical ectodermal ridge in the marsupial Monodelphis domestica. *Anat. Rec.*, (Hoboken). 2010 Aug;293(8):1325-32.

[140] Keyte AL, Smith KK. Developmental origins of precocial forelimbs in marsupial neonates. *Development,* (Cambridge, England). 2010 Dec;137(24):4283-94.

[141] Chew KY, Yu H, Pask AJ, Shaw G, Renfree MB. HOXA13 and HOXD13 expression during development of the syndactylous digits in the marsupial Macropus eugenii. *BMC developmental biology,* 2012;12:2.

[142] Behringer RR, Rasweiler JJt, Chen CH, Cretekos CJ. Genetic regulation of mammalian diversity. *Cold Spring Harb. Symp. Quant. Biol.*, 2009;74:297-302.

[143] Wilkie AO. Why study human limb malformations? *Journal of anatomy,* 2003 Jan;202(1):27-35.

[144] Farooq M, Troelsen JT, Boyd M, Eiberg H, Hansen L, Hussain MS, et al. Preaxial polydactyly/triphalangeal thumb is associated with changed transcription factor-binding affinity in a family with a novel point mutation in the long-range cis-regulatory element ZRS. *Eur. J. Hum. Genet.*, 2010 Jun;18(6):733-6.

[145] Wieczorek D, Pawlik B, Li Y, Akarsu NA, Caliebe A, May KJ, et al. A specific mutation in the distant sonic hedgehog (SHH) cis-regulator (ZRS) causes Werner mesomelic syndrome (WMS) while complete ZRS duplications underlie Haas type polysyndactyly and preaxial polydactyly (PPD) with or without triphalangeal thumb. *Human. mutation.*, 2010 Jan;31(1):81-9.

[146] Garrido-Allepuz C, Haro E, Gonzalez-Lamuno D, Martinez-Frias ML, Bertocchini F, Ros MA. A clinical and experimental overview of sirenomelia: insight into the mechanisms of congenital limb malformations. *Dis. Model. Mech.*, 2011 May;4(3):289-99.

[147] Alvarado DM, McCall K, Aferol H, Silva MJ, Garbow JR, Spees WM, et al. Pitx1 haploinsufficiency causes clubfoot in humans and a clubfoot-like phenotype in mice. *Human molecular. Genetics,* 2011 Oct 15;20(20):3943-52.

[148] Gurnett CA, Alaee F, Kruse LM, Desruisseau DM, Hecht JT, Wise CA, et al. Asymmetric lower-limb malformations in individuals with

homeobox PITX1 gene mutation. *American journal of human genetics,* 2008 Nov;83(5):616-22.

[149] Klopocki E, Kahler C, Foulds N, Shah H, Joseph B, Vogel H, et al. Deletions in PITX1 cause a spectrum of lower-limb malformations including mirror-image polydactyly. *Eur. J. Hum. Genet.,* 2012 Jun;20(6):705-8.

[150] Tanaka EM, Reddien PW. The cellular basis for animal regeneration. *Developmental cell,* 2011 Jul 19;21(1):172-85.

[151] Bryant SV, Endo T, Gardiner DM. Vertebrate limb regeneration and the origin of limb stem cells. *The International journal of developmental biology,.* 2002;46(7):887-96.

[152] Poss KD, Keating MT, Nechiporuk A. Tales of regeneration in zebrafish. *Dev. Dyn.,* 2003 Feb;226(2):202-10.

[153] Satoh A, Makanae A, Wada N. The apical ectodermal ridge (AER) can be re-induced by wounding, wnt-2b, and fgf-10 in the chicken limb bud. *Developmental biology,* 2010 Jun 15;342(2):157-68.

[154] Kawakami Y, Rodriguez Esteban C, Raya M, Kawakami H, Marti M, Dubova I, et al. Wnt/beta-catenin signaling regulates vertebrate limb regeneration. *Genes and development,* 2006 Dec 1;20(23):3232-7.

[155] Yokoyama H, Ide H, Tamura K. FGF-10 stimulates limb regeneration ability in Xenopus laevis. *Developmental biology,* 2001 May 1;233(1):72-9.

[156] Chan WY, Lee KK, Tam PP. Regenerative capacity of forelimb buds after amputation in mouse embryos at the early-organogenesis stage. *J. Exp. Zool.,* 1991 Oct;260(1):74-83.

[157] Reginelli AD, Wang YQ, Sassoon D, Muneoka K. Digit tip regeneration correlates with regions of Msx1 (Hox 7) expression in fetal and newborn mice. *Development,* (Cambridge, England). 1995 Apr;121(4):1065-76.

[158] Wanek N, Muneoka K, Bryant SV. Evidence for regulation following amputation and tissue grafting in the developing mouse limb. *J. Exp. Zool.,* 1989 Jan;249(1):55-61.

[159] Han M, Yang X, Farrington JE, Muneoka K. Digit regeneration is regulated by Msx1 and BMP4 in fetal mice. *Development,* (Cambridge, England). 2003 Nov;130(21):5123-32.

[160] Yu L, Han M, Yan M, Lee EC, Lee J, Muneoka K. BMP signaling induces digit regeneration in neonatal mice. *Development,* (Cambridge, England). 2010 Feb;137(4):551-9.

[161] Lehoczky JA, Robert B, Tabin CJ. Mouse digit tip regeneration is mediated by fate-restricted progenitor cells. Proceedings of the National

Academy of Sciences of the United States of America. 2011 Dec 20;108(51):20609-14.
[162] Hamburger V, Hamilton HL. A series of normal stages in the development of the chick embryo. *J. Morph.*, 1951;88:49-92.

In: Embryo Development
Editors: D. Reyes and A. Casales

ISBN: 978-1-62417-723-1
© 2013 Nova Science Publishers, Inc.

Chapter 4

THE USE OF TIME LAPSE PHOTOGRAPHY IN AN *IN VITRO* FERTILIZATION PROGRAMME FOR BETTER SELECTION FOR EMBRYO TRANSFER

Borut Kovačič, Nina Hojnik and Veljko Vlaisavljević

Department of Reproductive Medicine and Gynaecologic Endocrinology,
University Medical Centre Maribor,
Maribor, Slovenia

ABSTRACT

The time lapse photography is not a new method for assessing the dynamics of early embryo development *in vitro*. It has been used many times in the past for studying cleavages and blastulation of embryos of various animal species. However, this technique became available for routine use in an human *in vitro* fertilization (IVF) programme only a couple years ago and it becomes more and more popular today. The new time lapse systems are using modified microscopes which are positioned within the incubators. The observation of embryos does not need the opening of incubators. By sequential photographing of each embryo separately with camera of low intensity illumination, more than 1400 pictures of embryo are made. All these pictures are collected together and transformed into a short movie with software. This system offers the observation of dynamics of embryo development. The studies, which have used a time lapse technique for studying embryo develop- ment, revealed that the timing between different events can be used for predicting its developmental potential. In this paper the advantages and drawbacks of time lapse photography is precisely described. An overview through the published papers analyzing the dynamics of human embryo development from the zygote toward blastocyst is done and new timing parameters for grading zygotes, early embryos and blastocysts are analyzed.

INTRODUCTION

The overall success of *in vitro* fertilization (IVF) technique is despite tremendous progression since the beginnings staying relatively low, around 30% [1]. Much attention is given to optimize cultivation conditions for embryos in the laboratory. We can witness huge advances of the insemination techniques by introducing intracytoplasmic sperm injection (ICSI) [2]. The cultivation media have been improved and embryos can be cultured to the blastocyst stage. Incubators were modified and more physiological atmosphere with reduced oxygen concentration is being used [3]. Besides, most of the material used in IVF laboratories, is now tested for embryotoxicity with the mouse embryo assay (MEA).

The final step of *in vitro* cultivation, that is selection of the best embryo for embryotransfer, is mainly based on embryo morphology. With implementation of elective single embryo transfer (eSET) policy, to avoid complications due to multiple pregnancy, the selection of the most competent embryo is even of greater importance. For this reason, the attention is focused towards searching for new criteria for selection the embryo with best implanting potential.

There is a need to find new non-invasive methods for evaluating embryo quality. One of the possible method is the use of the time lapse photography technique which enables us to monitor closely the events in embryo development during *in vitro* cultivation. It has been used for research purposes on animal models for a long time [4, 5]. The first known study on human embryos has been also described early [6]. But only recently, with development of new equipment, which provides stable culture conditions for the embryos, it is available for the use in routine in IVF laboratories.

The underlaying biological mechanisms of the early embryo development are not known enough. These events are complex. The egg cell must complete the second female meiosis first, the sperm content must go through decondensation, the genetic material is modified through epigenetic reprogramming. The events that can be witness are the formation of male and female pronuclei, disappearance of the pronuclei, formation of the cleavage furrow, first cytokinesis, sequential divisions, cytoplasmic waves, formation of cytoplasmic fragments, compaction, blastulation, pulsations of blastocoel, hatching. The mitotic divisions are crucial for development of certain cell mass in a limited time frame when implantation is possible, before menstrual shredding of the endometrium begins. For successful development, timing of these events are very important. With time lapse imaging of *in vitro* cultivated embryos, we can witness these events that are usually entirely missed with static daily evaluation of embryo morphology in a given time point.

The main aim of using time lapse technology in IVF programme is to find those characteristics of early stages of embryo development that could predict the potential of an embryo to implant and lead to successful pregnancy. That is to find new dynamic markers of embryo quality that could improve success of IVF programme.

EMBRYO SELECTION

Evaluation of embryo morphology is the main method for selection prior embryotransfer in majority of IVF laboratories. The issue is being extensively studied and many morphologic characteristics of embryos have been shown to be good predictors for their further development [7, 8, 9, 10, 11, 12, 13].

Although there is close correlation between embryo morphology in given time points and it's ability to implant [14, 15, 16], with using only morphological criteria, we can still poorly predict if the embryotransfer will result in successful pregnancy.

There are many grading systems for the embryos in pronuclear stage, cleavage stage embryos, blastocysts and also scoring systems that combine several scoring parameters [17].

Blastocyst transfer is routinely used in many laboratories and it leads to better pregnancy rates than cleavage embryo transfers [18]. By extending cultivation to day 5, we propose that the selection is better. The embryos already begin their genomic activation and they go through some sort of natural selection, because many non-competent embryos will already arrest by this time. Blastocyst morphology is correlated with implantation ability [14, 16]. But there is some evidence that extended *in vitro* cultivation can have negative effect on embryos [19]. There is also evidence that culture conditions can affect embryo morphology [20]. Shorter cultivation means shorter exposure to possible non-optimal conditions that could affect embryo metabolism and epigenetic imprinting. The profound insight of the effect of media composition and cultivation conditions on various metabolic processes in the embryo is not yet possible. Recent studies suggest that significant effect of media can be observed in newborn [21].

Freezing of the supernumerary embryos enables utilization of more embryos from one stimulated cycle in more additional frozen-thawed cycles. And we might accept policy to try to utilize as many embryos as possible and transfer everything that is viable in several attempts. Vitrification as a superior method of cryopreservation offers good solution, but it is additional stress to the embryo. Not all embryos survive and implantation rates are lower than in fresh cycles [22]. This strategy exposes patients to big psychological stress because some of them have to go through many unsuccessful transfers or even abortions before delivery. Cryopreserving everything that is viable also raises problems with piling of material in cryobanks.

Grading of embryos on the ground of their morphological appearance is also submitted to certain subjectivity, interobserver variability and intraobserver reproducibility must be considered [23].

TIME LAPSE TECHNIQUE, IT'S USE AND SAFETY

The Technique

Time-lapse photography is a technique by which some events that are happening imperceptibly slowly are filmed in such a way, that when watching a film they appear much faster. It's being used for a century in film industry, especially in the documentary films. Classical scenes filmed in this way are plant growth, daily traffic in the city or movement of clouds and stars in the sky.

Usually films are made by taking 24 pictures (frames) per second and also projected in this way. Using time lapse technique, less pictures are taken, for example one every 10 minutes, but then these pictures are merged into a film that is perceptible to human eye.

Human embryos seem static when observed once a day for evaluating their temporal morphology. Many important events are missed entirely by such observation. Therefore time lapse technology fascinates embryologists because embryos are seen in quite a different perspective. The dynamics of the embryo development appears much more complex. Only in such a way some processes that are too slow to be noticed become obvious to us. All events that happen during cultivation *in vitro* can be projected in a 30 second movie. It enables us to witness events that are not well explained, like the direct division from one cell stage to three cell stage, formation and re-absorption of the fragments or even fusion of the cells and thus reducing number of cells in time.

The idea to detect events in early embryo development *in vitro* is not new. In a study of the time course of fertilization of human oocytes with intracytoplasmic sperm injection (ICSI), oocytes were observed every 2 hours post fertilization [24].

More frequent acquisition of images is needed to provide images of all important events, but in the meantime safety of the embryos must be considered. Exposing the embryos to the conditions outside the incubator and the microscopic light can have detrimental effects on embryo development [25]. Therefore suitable equipment was needed that would enable image acquisition in suitable and stable conditions for the embryos.

The Equipment

Payne et al. conducted time lapse recording on the stage of the inverted microscope with Nomarski optics with special environmental chamber which maintained 37°C, 5%CO_2 [6]. By the switching box the microscope lamp was turned on every minute for 5 seconds and in meantime video recorder (S-VHS time-lapse video cassette recorder, AG-6730) captured one frame. The camera was low-light CCD (charge-coupled device) colour video camera. The recording period was 17-20 h. The videos were viewed on monitor (final magnification x1064) and timing of the events relative to sperm injection was analysed.

Similar type of equipment, that was not commercially available, but was rather built in the laboratory, has also used in research work in more recent articles [26, 27]. The work of Wong et al. (2010) also presented modified microscopes with white-light LED illumination and custom built miniature microscopes that were placed in the standard incubator. They have used dark-field illumination and high sensitivity camera sensors.

Recently new time lapse equipment is available that has been developed for use in IVF laboratories.

Primo Vision system (Cryo-Innovation, Budapest, Hungary) is a compact microscope of 220x80x110 mm that is placed in a standard incubator.

Recently released version has improved optics with Hoffman modulation contrast. The microscope has green led light source and CCD camera (2560x 1920 pixels). WOW (well-of-the-well) dishes used for embryo cultivation have 9 or 16 microwells under one droplet of cultivation media covered with paraffin layer. The dish is placed on a holder under which is objective which is mechanically focused. The controlling unit is built for maximum 6 microscopes, thus maximum of 6 patients and 96 embryos. Time intervals between images can be adjusted from up to 5 minutes onwards.

Figure 1. Primo Vision (Cryo-Innovation, Budapest, Hungary) compact microscope is placed in the standard incubator.

A lot of published work has been done by using EmbryoScope™ time-lapse system (Unisense Fertilitech, Denmark). This is an integrated microscope and tri-gas incubator built to minimize disturbance of the culture conditions while the high resolution images are taken within controlled environment.

Six disposable EmbryoSlide® culture dishes holding 12 embryos each can be placed in the chamber. Due to small chamber the incubator allows fast regulation of CO_2 and O_2. When replacing culture medium, the changes are also minimized by the fact, that individual slides can be removed from the incubator independently. The Leica optics with Hoffman modulation contrast enables high quality imaging, it uses high resolution (1280 x 1024 pixels) camera and red LED light of 635 nm wavelength, enables multiple focal planes images in up to 20 min intervals.

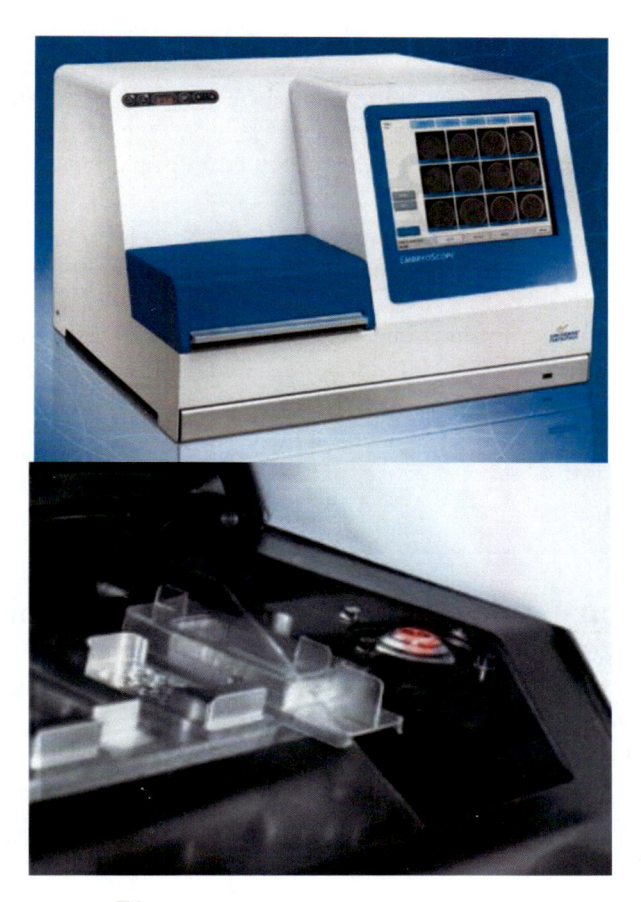

Figure 2. EmbryoScope™ (Unisense Fertilitech, Denmark); left: the whole incubator with integrated microscope; right: the movable tray positions each patient's culture dish above the objective when imiges are taken.

Japanese Sanyo developed an integrated system of a 50 l incubator with in-built microscope and moving stage. It has Olympus phase contrast or relief-contrast optics. The illumination is white led light, CCD camera of 790.000 pixels. It can be used for 12 patients, each maximum 10 embryos.

Tips for Implementation of Time Lapse Technology in IVF Laboratory

Firstly, the selection of time lapse system must be done according to specific needs and planned use of the equipment. The parameters that are important for the selection of the suitable system are the minimal time interval between each image acquisition, number of embryos that can be simultaneously monitored, number of focal planes, type of optics. There are also differences in dishes used for embryoculture in time lapse system, some enable co-cultivation.

Technical training prior the use is advisable, since preparation of the media must be carried out very carefully to avoid bubbles formation that could diminish image quality.

If the clinic wish to find some additional parameters that could improve embryo selection, careful planning of the study and sufficient number of embryos must be evaluated. The data from other centres cannot be simply implemented in another environment.

SAFETY

The gradual development of suitable time lapse equipment for human embryos had to overcome the problem of changing stable culture environment while taking images. Frequent evaluation of embryos outside the incubator can have detrimental effect on their implanting potential [25].

The periodic exposure to light during time-lapse observation might also present a risk to the embryo. The possible detrimental effect of light has been studied on hamster embryos [28]. There might be species specific sensitivity to light and different effect of various wavlenghts [29].

Embryos are exposed to quite an amount of light during *in vitro* cultivation, the total exposure has been calculated [30].

Therefore, before implementing time lapse technology from research purpose to the clinical routine, extensive testing has to be done to prove safety of the equipment. There are strict standards for introducing new technology in IVF laboratories [31].

Due to ethical issues linked to research on human embryos and the fact that possible negative effects might become evident many years after the child has been born, there are specific problems in this field of medicine.

Animal models were used to test the safety of the equipment [32]. Several authors reported data on safety from the results on donated embryos [27, 33].

The majority of the articles present data on safety of technology as a supplementary data while already conducting images acquisition [34]. Recently there are few studies with primary focus on the safety of the technique [35, 36, 37].

Table 1. Data on safety of the use of different time lapse equipment in IVF programme

Study		Subject (Control)	Equipment	Measures	Results
Cruz et al., 2011	Human	238 embryos – TL (240 control)	EmbryoScope 20 min, 7 focal planes	Embryo morphology, blastocyst morphology, embryo utilization, pregnancy rate	No significant difference with control for all evaluated parameters
Nakahara et al., 2010	Human	84 fertilized oocytes – TL (84 control)	Sanyo MCOK-5M 15 min, 3 focal planes	Fertilization rate, embryo morphology day2, day3	No significant difference with control for all evaluated parameters
Kirkegaard et al., 2012	Human RTC	297 oocytes – TL (303 control)	EmbryoScope no image acquisition, just testing equipment as an incubator	Proportions of 4-cell embryos on day 2, 7-8 cell embryos on day 3, blastocysts on day 5, implantation	No significant difference with control for all evaluated parameters
Payne et al., 1997	Human	Single randomly selected oocyte from each patient (102 fertilized oocytes) – TL (487 control)	Custom made 5 min	Fertilization, embryo utilization	No significant difference with control for all evaluated parameters

TL- time lapse, RTC-randomized clinical trial.

THE MAIN EVENTS IN EARLY EMBRYO DEVELOPMENT EVALUATED BY TIME LAPSE PHOTOGRAPHY

Time lapse technology is useful for observations of various phenomena in an early embryo development. The timing of the phases in early cleavage embryos can be measured and other features like multinucleation, cleavage furrow formation, fragmentation, cytoplasmic waves can be observed. Later the compaction of the embryo into the morula can be viewed, the formation of the blastocoel and it's pulsations, formation of the inner cell mass and hatching.

Different authors presented the time intervals between various stages in quite a different way. We must know the difference between cleavage, which is actually duplication of the cells (one- to two-, two- to four-cell stage embryo ect.) and mitotic divisions (one- to two-cell stage embryo, two- to three-cell stage embryo ect.).

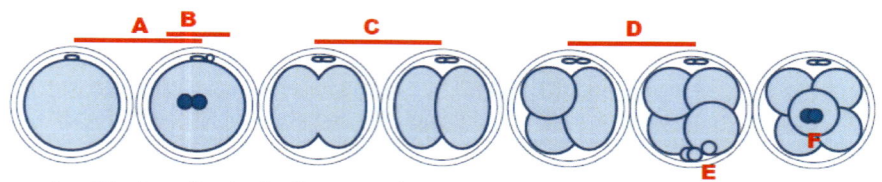

A. Timing of polar bodies extrusions and pronuclei appearence.
B. Synchronicity of male and female pronucleus formation.
C. Duration of the first blastomere division (cytokinesis).
D. Synchronicity of 2nd and 3rd mitoses (second embryo cleavage).
E. Fragments appearance and disappearance.
F. Multinucleation.

Figure 3. Graphical representation of some of the observed events and features.

Extrusion of Second Polar Body

The extrusion of the second polar body (PB) marks the completion of two meiotic divisions. Human oocytes enter meiosis I in early female fetal life. In diplotene stage the recombination of genetic material occurs in the process of crossing over, the oocytes are then arrested in the dictyate stage of prophase I. They stay in this form for up to decades.

Table 2. Presentation of studies of using kinetic markers as quality markers of embryo development

Observed parameter	Study	Subject	Overall mean time (±SD or min, max)	Results
Extrusion of the 2nd polar body	Payne et al., 1997	37 fertilized oocytes	2:39 (4:24, 8:00) post fertilization	Good quality day-3 embryos (n=17) 2:45 ± 0:59 Poor quality embryos (n=13) 2:04 ± 1:04 (P=0,03)
	Mio, 2006	55 fertilized oocytes	2:00 ±1:00 post fertilization	No comparison
	Azzarello et al., 2012	159 transfered embryos (46 resulted in live birth)	/	Zygotes resulted in live birth (n=46) 3:47 ± 0:17 Zygotes not resulted in live birth (n=113) 3:37±0:11 (NS
Formation of female and male pronucleus and their disappearance	Payne et al., 1997	37 fertilized oocytes	4:59 (2:51, 9:11) Injection to male PN 4:59 (2:51, 11:26) Injection to female PN	Synchronicity of male, female PN Good quality day-3 embryos (n=17) 0:11 ± 0:22 Poor quality day-3 embryos (n=13) 0:12 ± 0:37 (P=0,03)
	Mio, 2006	55 fertilized oocytes	5,8 ±2,5 h Injection to male PN 5,9±2,5h Injection to female PN	No comparison
	Lemmen et al., 2008	102 fertilized oocytes	/	Time of disappearance of the pronuclei correlates with blastomere number on day 2 (P=0,0016)
	Azzarello et al., 2012	159 transfered embryos (46 resulted in live birth)	/	Time of disappearance of the pronuclei: Zygotes resulted in live birth (n=46) 24.52 ± 0:35 Zygotes not resulted in live birth (n=113) 23:10±0:23 (P=0,022)
First cleavage	Mio, 2006	55 fertilized oocytes	24,8 ±4,2 h Injection to initiation of the first cleavage	No comparison
	Lemmen et al., 2008	102 fertilized oocytes	/	Time of the initiation of the first cleavage significantly correlates with blastomere number on day 2 (P=0,0014)
	Meseguer et al., 2011	247 embryos with known implantation	/	Implanted embryos (n=61) 25,6±2,2h Not implanted embryos (n=186) 26,7±3,8h (P=0,022)

Table 2. (Continued)

Observed parameter	Study	Subject	Overall mean time (±SD or min, max)	Results
Duration of 1st cytokinesis	Mio, 2006	55 fertilized oocytes	Duration of the first cytokinesis median 0,5 h	No comparison
	Wong et al, 2010	100 embryos	/	Duration of the first cytokinesis in embryos that developed to the blastocysts median 14,3±6,0 min
Second cleavage	Wong et al, 2010	100 embryos	/	Time interval from the end of the first mitosis to the initiation of the second one embryos that developed to the blastocysts median 11,1±2,2 h
Observed parameter	Study	Subject	Overall mean time (±SD or min, max)	Results
	Meseguer et al., 2011	247 embryos with known implantation	/	Time of division to 3-cells (t3): Implanted embryos (n=61) 37,4±2,8h Not implanted embryos (n=185) 38,4±5,2 (P=0,002)
	Hlinka et al., 2012	28 implanted embryos	/	Duration of the 2-cell stage: 11±1 h
Synchronicity of the 2nd and 3rd mitoses	Wong et al., 2010	100 embryos	/	Time interval between the 2nd and the 3rd mitoses in embryos that developed to the blastocysts: 1,0±1,6 h
	Meseguer et al., 2011	247 embryos with known implantation	/	Duration of transition from 2- to 4-cell embryo (s2) Implanted embryos (n=61) 0,78±0,73h Not implanted embryos (n=182) 1,77±2,83 (P=0,016)
Division to 5 cell-embryo	Meseguer et al., 2011	247 embryos with known implantation	/	Duration from injection to 5-cell embryo (t5) Implanted embryos (n=61) 52,3±4,2h Not implanted embryos (n=167) 52,6±6,8 (P<0,001)

In response to LH (luteinizing hormone) surge, the oocyte resumes meiosis and the extrusion of the first PB marks the end of meiosis I. Then the second meiotic division begins and progresses to metaphase II stage, when it is arrested once again. The second arrest is required for the final cytoplasmic maturation.

The entry of spermatozoa is needed to proceed the meiosis II and the completion is marked by extrusion of the second PB.

The timing of second PB extrusion has been studied early after introducing ICSI. No correlation has been established between the time of extrusion and subsequent speed of development or embryo quality [38]. They observed 135 oocytes 1 hour, 2 hours and 3 hours post ICSI. 56.6% of oocytes extruded second PB within the first 2 h and 78.3% 3 h after injection.

Kinetics of second PB extrusion observed with time lapse technique and relation with the further embryo development is described in several works [6, 39, 40]. Payne et al. observed 30 oocytes and established that extrusion occurs earlier in good quality embryos (2h45min ± 59min in good vs. 3h23min ± 1h40min in poor quality embryos) [6]. When comparing the embryos resulting in live birth with those that did not, no significant difference has been found in the time interval from fertilization to the extrusion of the second PB [40].

There is not enough data so far for using timing of extrusion of the second PB as a quality marker of the oocyte, or perhaps even quality marker of the sperm, as it might reflect decondensation ability of spermatozoa. It is also not enough data that this kinetic parameter could be used as quality marker of embryo development.

Pronuclear Formation and Breakdown

Among different parameters suggested for efficient embryo scoring system, pronuclear morphology has been extensively studied. While some authors proved their prognostic value [41] others did not find any importance [42].

Time lapse observations of pronuclear appearance, disappearance, size and number of the nuclear precursor bodies (NPB) interestingly revealed that these are dynamic features that change in time [40]. The Z scoring for the NPB changes as the they go through a development process, which happens in a short time. One static observation in between the time interval 16 to 18 hours post insemination can therefore be inappropriate.

The timing of appearance of the female and male pronucleus and their later disappearance has also been studied [6, 34]. In the study of Lemmen et al. the embryos that developed in 4-cell embryos on day two of cultivation showed significantly earlier disappearance of the pronuclei than 2-cell embryos [34].

The Cleavages

The Timing

At the time of fertilization fully differentiated oocyte and sperm cell must go through reprogramming toward quick divisions and totipotency of the daughter cells. The timing of cell divisions are governed by cell cycle regulators, that are quite unique in the 8-cell embryo and might be markers of totipotency [43]. Based on this study, which uses microarray analysis of the whole human genome, the authors suggest that the usual checkpoints are not active in totipotent blastomers and that it is possible that trancriptional and translational feedback loops which form the cirkadian clocks are the basis of early embryo development.

Nevertheless the cells in the cleaving embryo must go through usual phases of the cell cycle at certain time frame to establish the critical mass of cells that can maintain the pregnancy.

It has been established earlier that timing of the first cleavage can be used as a marker of embryo quality [44, 45].

With time lapse technique the process can be viewed in a more comprehensive way. The timing of the first and subsequent cleavages was studied on mice embryos [46] and the conclusion was that the first cleavage is very important, embryos with slow first cleavage showed inferior developmental potential. Mice model was also used in the work by Pribenszky et al., (2010) [32] that showed importance of the timing of the first cleavage for the blastocyst development.

The timing of sequential divisions are also predictive for blastocyst development. Wong et al. (2010) [27] showed that interval between second and third division is predictive for development to blastocyst. Duration of cytokinesis significantly correlated with the pregnancy [47].

With time lapse imaging, not only the timing, but also the form of the cytoplasmic furrow, that is the first step of cytokinesis, can be studied in details.

The Cytokinesis

Cytokinesis is mechanical division of the mother cell into two daughter cells and can be observed using time lapse technique. The process is governed by biochemical signals, cell shape and cellular mechanics. The site of the division must be carefully positioned regard to the chromosomes, so that each daughter cell receives a single copy of the genome.

Protein kinases and signalling pathways direct the process. The mechanical force is generated by the actomyosin-based equatorial contractile ring. There has also been proposed that cell surface actin elements generate cortical mechanic properties and that local stiffness or softness of the cell surface has important role in cytokinesis [48]. Recently it has been established that the stiffness is reduced in equatorial region during cytokinesis, but remains in the polar region [49]. The process is governed by interaction of ZEN-4 protein, which is the component of the centralspindlin and GTPase Rac which has a role in actin meshwork under the cell surface. In the experiments on *C. elegans* embryo model, the equatorial reduction of stiffness alone was sufficient for the formation of the cleavage furrow, without the action of the contractile ring [49].

The abnormalities of cytokinesis can be visualised using time lapse technique. Bipolar cytoplasmic furrow, clear separation of the daughter cells withouth formation of the fragments and the narrow time interval between the appearance of the cytoplasmic furrow and complete separation of the daughter cells, are the markers of successful cytokinesis. Other patterns (Figure 5) also appear which are abnormal. The unipolar furrow is ingression only on the one side of the cell and usually leads to extensive formation of the fragments. Multiple cleavage furrows lead to membrane ruffling, prolonged cytokinesis and fragmentation. The tripolar furrow leads to direct cleavage from one to three cells, usually without fragmentation. The furrow is important for duration of the cytokinesis and it has been showed that the duration of the first cytokinesis is predictive for blastocyst development [27].

Kinetic Markers Linked to Gene Expression and Chromosomal Status

The remarkable work of Wong et. al, (2010) [27] correlated kinetic parameters of embryo development with transcriptional patterns studied with qRT-PCR analysis. They analysed 9 genes assumed to have important role in cytokinesis. The gene expression profiles in arrested embryos differ from those that developed normally.

They have concluded that successful development to blastocyst stage is to some extent determined very early in embryo development, before the embryonic gene activation occurs. Their data explain that the fate of an embryo depends strongly on factors inherited from the oocyte. It can be assumed that transcripts in oocytes regulate first divisions that are crucial for developmet to blastocyst. That is why are these early kinetic markers so predictive for blastocyst development, although the effect of sperm factors and genetic causes that could alter the outcome must also be considered.

There are some studies that tried to link kinetic parameters to the ploidy of the embryos. Some scarce data suggested that aberrant divisions (direct division from 2- to 5-cell stage) are correlated with complex aneuplo- idy and that time interval between 3-cell stage and 4-cell stage is significantly correlated with the karyotype of the embryo [50].

Early morphokinetic data can be very predictive for blastocyst development. Wong et al., showed that correct timing of first cytokinesis, time interval between first and second mitoses and time interval between second and third mitoses predict blastocyst development with 94% sensitivity and 93% specificy. But as there are no data on implantation, it can be presumed that predictive capacity of kinetic parameters for the implantation is lowered due to sperm related factors, ploidy status and genetic defects.

Embryo Selection

Apart from basic research done with this technology, introducing time-lapse recording of *in vitro* cultivated embryos in daily routine in an IVF laboratory might enable better selection among many for the embryotransfer.

The group of Meseguer et al. developed a multivariable model based on many time lapse recordings which enables to classify embryos according to their probability to implant [47].

They have monitored embryos of 285 couples (standard IVF and donation programme) in EmbryoScope system for three days. 522 embryos were transferred, for 247 they have obtained full information about their implantation. They proposed the scheme of analysed time variables:

- $t2$ = time to cleavage to 2-cell embryo
- $t3$ = time to cleavage to 3-cell embryo

- t4 = time to cleavage to 4-cell embryo
- t5 = time to cleavage to 5-cell embryo
- cc2 = second cell cycle (t3-t2)
- s2 = synchrony of divisions leading to 3 and 4 cell embryo (t4-t3)

Based on the data they made a classification tree to divide embryos in 10 categories. The classification was based on both kinetic markers and morphological markers.

They found that the parameters which significantly correlated with higher implantation are: a) time of division to 5 cells, b) time between division to 3 cells and division to 4 cells and c) time between division from 2 and division to 3 cells.

Recently they published data from a retrospective study on a large number of cycles (1390 time lapse, 5915 standard incubator) and showed the difference of 20% in pregnancy rates in favour of using time lapse technology [51]. But the study could not deduce that the improvement is based on new kinetic markers used for embryo selection.

Another group reported that early embryo cleavage timing can be the useful selection criteria when the selection between blastocysts has to be done, it can narrow the group of blastocyst with high probability of implanting [52]. They have classified embryos it two categories based on timing of events of the group of implanted embryos (28 implanted embryos).

There is no large prospective randomized clinical trial so far that would prove the advantages of using time lapse technology for embryo selection in human IVF programme.

EXPERIMENTAL PART

The time lapse observations were made on 73 embryos from 12 blastocyst cycles. The main goal was to detect the specific cleavage abnormalities and to observe their impact on development to blastocyst. The abnormalities of the first cytokinesis were unipolar, tripolar and multipolar cytoplasmic furrow.

Female patients (n=12) were down-regulated with GnRH agonists or antagonists, stimulated with gonadotrophines. Their mean age was 33,25 ± 5,0. We had 5 cases of male infertility, 2 female, 3 combined, 2 idiopathic. Mean number of oocytes retrieved at the punction was 7,17 ± 1,6, all were inseminated with ICSI method.

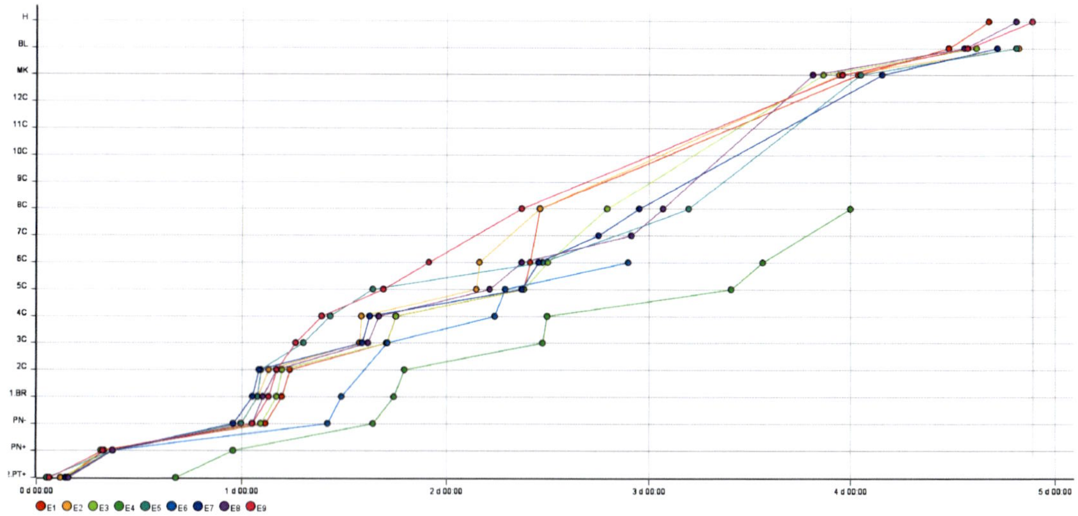

- 2.PT: extrusion of the second polar body
- PN+: appearance of the two pronuclei
- PN-: disappearance of the two pronuclei
- 1.BR: the formation of the first cytoplasmic furow
- MK: compaction into morula
- BL: expanded blastocyst
- H: hatching blastocyst
- 2C,3C,4C....: 2-cell stage, 3-cell stage, 4-cell stage

Figure 4. Graphical representation of the kinetic parametrs for nine embryos of one patient. On the ordinate are the stages of development and on the abscissa the time in hours.

We obtained 73 embryos that were cultivated for 5 days in sequential medium (Origio) in standard incubator (6% CO_2, 5%O_2). We classified blastocysts in the optimal, suboptimal group and the group of arrested embryos according to previously described blastocyst scoring system [16].

Primo Vision time lapse system has been used and image acquisition was made in intervals of 5 minutes in 7 focal planes for 5 days, light exposure was 20 ms. Timing of the events has been reviewed manually and for each patient graphical representation of the embryo kinetics has been plotted with the use of the software.

The median of time intervals are presented in the table. There are also presented frequencies of specific cleavage abnormalities (unipolar, tripolar and multipolar cleavage furrows). In the group of optimal blastocysts, no cleavage abnormalities were found. In the group of suboptimal blastocysts two cleavage abnormalities were found in the first cytokinesis and two in the 2-cell stage embryo. But in the group of arrested embryos the frequency increase (9 in 1-cell stage, 1 in 2-cell stage).

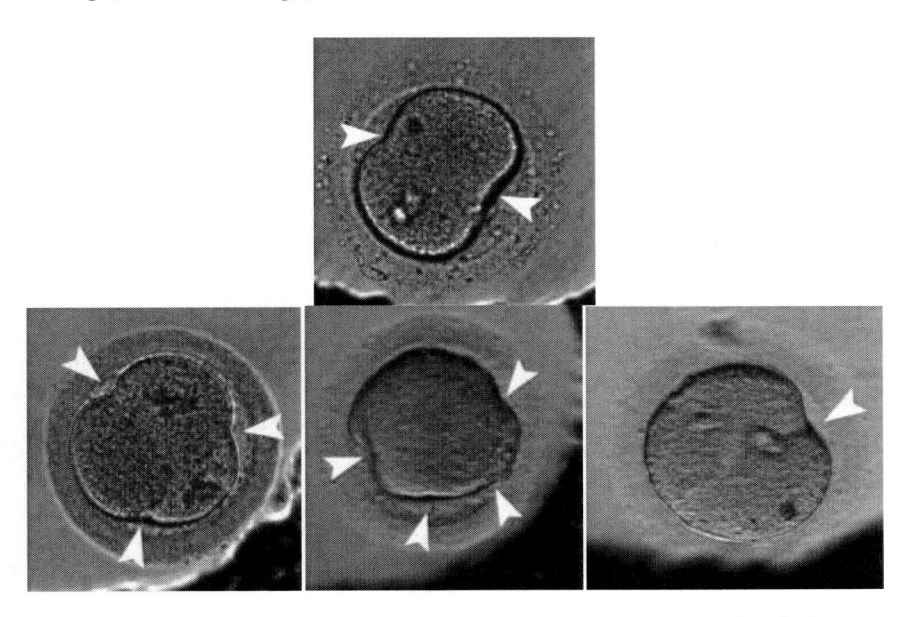

Figure 5. The upper image presents a zygote with normal bipolar cytoplasmic furrow. Arrows indicate sites of ingression. In the lower row on the left side is zygote with tripolar furrow, in the middle is zygote with multipolar furrows and on the right side a zygote with unipolar furrow.

Table 3. Median and standard deviation for time intervals of evaluated parameters in the groups of optimal blastocysts, suboptimal blastocysts and arrested embryos

		Optimal blastocysts	Suboptimal blastocyst	Arrested embryos
		n=18	n=19	n=36
F - 2nd PB	mean ± sd	3h 59 min ± 2h 25 min	3h 22 min ± 1h 43 min	3h 51 min ± 14h 36 min
	min	1h 20 min	1h 6 min	9 min
	max	12h	7h 8 min	11h 50 min
F - PN+	mean ± sd	9h 1 min ± 1h 36 min	9h 56 min ± 2h 47 min	10h 59 min ± 3h 16 min
	min	7h 1 min	7h 1 min	7h 1 min
	max	12h 24 min	16h 4 min	20h 10 min
F - PN-	mean ± sd	24h 37 min ± 2h 7 min	26h 36 min ± 1h 37 min	26h 48 min ± 4h 54 min
	min	21h 2 min	23h 2 min	21h 2 min
	max	28h 33 min	34h 48 min	40h 45 min
F - 1st CF	mean ± sd	26h 12 min ± 1h 59 min	28h 19 min ± 3h 17 min	29h 11 min ± 5h 59 min
	min	21h 57 min	24h 25 min	21h 57 min
	max	29h 28 min	36h 38 min	51h 4 min
ACF	n (%)	0 (0%)	2 (10,5%)	8 (22,2%)
	unipolar	n=0	n=1	n=0
	tripolar	n=0	n=0	n=5
	multipolar	n=0	n=1	n=3
F - 2c	mean ± sd	27h 20 min ± 1h 52 min	29h 23 min ± 2h 29 min	30h 27 min ± 4h 53 min
	min	23h 50 min	25h 27 min	23h 50 min
	max	31h 3 min	43h 13 min	43h 13 min
	eq.	n=14	n=15	n=24
	uneq.	n=4	n=4	n=5
cytokin.	mean ± sd	1h 8 min ±1h 4 min	1h 5 min ± 25min	2h 37 min ± 6h 3 min
	min	15 min	15 min	15 min
	max	4h 50 min	6h 35 min	35h 50 min

		Optimal blastocysts	Suboptimal blastocyst	Arrested embryos
			Abnormal cleavage	
4c-3c	mean ± sd	1h 18 min ± 2h 44 min	58 min ± 46 min	3h 19 min
	min	0	0	0
	max	10h 26 min	2h 16 min	28h 5 min
4c	mean ± sd	38h 53 min ± 2h 51 min	42h 41 min ± 4h 2 min	46h 23 min ± 8h 12 min
	min	33h 12 min	38h 4 min	33h 12 min
	max	45h 4 min	53h 53 min	74h 38 min
	eq.	n=15	n=13	n=17
	uneq.	n=3	n=5	n=10
8c	mean ± sd	62h 50 min ± 8h 3 min	67h 55 min ± 11h 47 min	71h 24 min ± 16h 5 min
	min	50h 45 min	53h 32 min	50h 45 min
	max	82h 58 min	90h 1 min	105h 38 min

F-2.PB: time between fertilization and extrusion of the second polar body.
F-PN+: time between fertilization and appearance of the pronuclei.
F-PN-: time between fertilization and di appearance of the pronuclei.
F- 1st CF: time between fertilization and appearance of the first cleavage furrow.
F-2c: time between fertilization and division to 2-cell stage.
cytokin: duration of the first cytokinesis.
4c-3c: time between 3-cell stage and division to 4-cell stage.
4c: time between fertilization and 4-cell stage.
8c: time between fertilization and 8-cell stage.
ACF: abnormal cleavage furrows
*: in the parenthesis are frequencies of direct divisions from 2-cell to 5-cell embryos.

The formation of normal cytoplasmic furrow is very important for successful division. When unipolar indentation occurs, the division is usually prolonged with intensive membrane ruffling and extensive formation of cytoplasmic fragments. The interesting finding is that tripolar indentation (in 5 embryos), which leads to direct division from 1-cell to 3-cell embryo, was observed in two patients with numerous previous unsuccessful IVF attempts. We observed abnormal cytoplasmic furrow pattern in 10 embryos. Two embryos with abnormal furrow reached the blastocyst stage, but were classified as suboptimal blastocysts. Another eight embryos with this kind of abnormalities were arrested before fifth day of development. It has been proven in previous studies that such embryos have very little or no chances to lead to successful pregnancy [47].

It is possible that these cleavage abnormalities reflect lack of some proteins involved in cytokinesis. The work of Wong et al., (2010) [27] clearly showed that expression of genes involved in cytokinesis and early cleavages are related to successful development to the blastocyst stage. Abnormalities reflect oocyte competence, it's ability for the spindle formation and correct organization of all the elements of cytoskeleton and underlying signaling pathways. The consecutive status of an embryo, that has not been able to form the spindle correctly, is uneven distribution of genetic material, thus aneuploidy. We can summarize that cleavage furrow pattern of first cytokinesis is one of very predictive embryo characteristics that could help in selection of the best early embryo for transfer.

Concerning the time intervals for the three subgroups we observed that all the intervals were prolonged in the group of arrested embryos. However, the standard deviations were so big in all three groups that these time intervals overlapped. While some authors claim that time intervals that predict successful embryo development are quite narrow [27], others found no significant difference in mean times between implanted and non-implanted group [47]. Our results are in agreement with the later study. The reasons for this discrepancy are still unknown and can lay in big variations in culture conditions between studies and in many unknown intrinsic factors within zygotes.

Other strategy for use in daily clinical practice would be direct comparison of time intervals of individual parameters prior selection for embryotransfer in case of many blastocyst of equal morphological grade on day five of cultivation. Some authors claim, that this additional information could improve the selection and increase the success rate of treatment [52].

CONCLUSION

Recent studies of kinetic parameters in relation to embryo quality are interesting new information for *in vitro* cultivation of human embryos, especially for the selection of the most competent embryo for the embryo-transfer.

Many studies have proved that there is correlation between several kinetic parameters and ability of embryo to develop to blastocyst or it's implantation capacity.

However, there is a need for large randomized clinical studies that would analyse the possible better selection on the ground of kinetic parameters in comparison with the standard morphological criteria.

REFERENCES

[1] Ferraretti, A. P., et al., Assisted reproductive technology in Europe, 2008: results generated from European registers by ESHRE. *Hum Reprod*, 2012. 27(9): p. 2571-84.

[2] Palermo, G., et al., Pregnancies after intracytoplasmic injection of single spermatozoon into an oocyte. *Lancet*, 1992. 340(8810): p. 17-8.

[3] Kovacic, B. and V. Vlaisavljevic, Influence of atmospheric versus reduced oxygen concentration on development of human blastocysts *in vitro*: a prospective study on sibling oocytes. *Reprod Biomed Online*, 2008. 17(2): p. 229-36.

[4] Bavister, B., A minichamber device for maintaining a constant carbon dioxide in ari atmosphere during prolonged culture of cells on the stage of an inverted microscope. *In Vitro Cellular & Developmental Biology - Plant*, 1988. 24(8): p. 759-763.

[5] Gonzales, D. S., Pinheiro J. C. and Bavister B. D., Prediction of the developmental potential of hamster embryos *in vitro* by precise timing of the third cell cycle. *J Reprod Fertil*, 1995. 105(1): p. 1-8.

[6] Payne, D. et al., Preliminary observations on polar body extrusion and pronuclear formation in human oocytes using time-lapse video cinematography. *Hum Reprod*, 1997. 12(3): p. 532-41.

[7] Steer, C. V. et al., The cumulative embryo score: a predictive embryo scoring technique to select the optimal number of embryos to transfer in

an in-vitro fertilization and embryo transfer programme. *Hum Reprod,* 1992. 7(1): p. 117-9.
[8] Desai, N. N. et al., Morphological evaluation of human embryos and derivation of an embryo quality scoring system specific for day 3 embryos: a preliminary study. *Hum Reprod,* 2000. 15(10): p. 2190-6.
[9] Rienzi, L. et al., Day 3 embryo transfer with combined evaluation at the pronuclear and cleavage stages compares favourably with day 5 blastocyst transfer. *Hum Reprod,* 2002. 17(7): p. 1852-5.
[10] Ebner, T. et al., Selection based on morphological assessment of oocytes and embryos at different stages of preimplantation development: a review. *Hum Reprod Update,* 2003. 9(3): p. 251-62.
[11] Scott, L. et al., Morphologic parameters of early cleavage-stage embryos that correlate with fetal development and delivery: prospective and applied data for increased pregnancy rates. *Hum Reprod,* 2007. 22 (1): p. 230-40.
[12] Finn, A. et al., Sequential embryo scoring as a predictor of aneuploidy in poor-prognosis patients. *Reprod Biomed Online,* 2010. 21(3): p. 381-90.
[13] Montag, M., Liebenthron J. and Koster M., Which morphological scoring system is relevant in human embryo development? *Placenta,* 2011. 32 *Suppl* 3: p. S252-6.
[14] Gardner, D. K. et al., Blastocyst score affects implantation and pregnancy outcome: towards a single blastocyst transfer. *Fertil Steril,* 2000. 73(6): p. 1155-8.
[15] de Neubourg, D. et al., Single top quality embryo transfer as a model for prediction of early pregnancy outcome. *Hum Reprod,* 2004. 19(6): p. 1476-9.
[16] Kovacic, B. et al., Developmental capacity of different morphological types of day 5 human morulae and blastocysts. *Reprod Biomed Online,* 2004. 8(6): p. 687-94.
[17] The Istanbul consensus workshop on embryo assessment: proceedings of an expert meeting. *Hum Reprod,* 2011. 26(6): p. 1270-83.
[18] Blake, D. A. et al., Cleavage stage versus blastocyst stage embryo transfer in assisted conception. *Cochrane Database Syst Rev,* 2007(4): p. CD002118.
[19] Kallen, B. et al., Blastocyst versus cleavage stage transfer in *in vitro* fertilization: differences in neonatal outcome? *Fertil Steril,* 2010. 94(5): p. 1680-3.

[20] Kiessling, A. A. et al., Development and DNA polymerase activities in cultured preimplantation mouse embryos: comparison with embryos developed *in vivo*. *J Exp Zool*, 1991. 258(1): p. 34-47.
[21] Dumoulin, J. C. et al., Effect of *in vitro* culture of human embryos on birthweight of newborns. *Hum Reprod*, 2010. 25(3): p. 605-12.
[22] van Landuyt, L. et al., Outcome of closed blastocyst vitrification in relation to blastocyst quality: evaluation of 759 warming cycles in a single-embryo transfer policy. *Hum Reprod*, 2011. 26(3): p. 527-34.
[23] Baxter Bendus, A. E. et al., Interobserver and intraobserver variation in day 3 embryo grading. *Fertil Steril*, 2006. 86(6): p. 1608-15.
[24] Nagy, Z. P. et al., Time-course of oocyte activation, pronucleus formation and cleavage in human oocytes fertilized by intracytoplasmic sperm injection. *Hum Reprod*, 1994. 9(9): p. 1743-8.
[25] Zhang, J. Q. et al., Reduction in exposure of human embryos outside the incubator enhances embryo quality and blastulation rate. *Reprod Biomed Online*, 2010. 20(4): p. 510-5.
[26] Mio, Y. and Maeda K., Time-lapse cinematography of dynamic changes occurring during *in vitro* development of human embryos. *Am J Obstet Gynecol*, 2008. 199(6): p. 660 e1-5.
[27] Wong, C. C. et al., Non-invasive imaging of human embryos before embryonic genome activation predicts development to the blastocyst stage. *Nat Biotechnol*, 2010. 28(10): p. 1115-21.
[28] Takahashi, M. et al., Assessment of DNA damage in individual hamster embryos by comet assay. *Mol Reprod Dev*, 1999. 54(1): p. 1-7.
[29] Takenaka, M., Horiuchi T. and Yanagimachi R., Effects of light on development of mammalian zygotes. *Proc Natl Acad Sci USA*, 2007. 104(36): p. 14289-93.
[30] Ottosen, L. D., Hindkjaer J. and Ingerslev J., Light exposure of the ovum and preimplantation embryo during ART procedures. *J Assist Reprod Genet*, 2007. 24(2-3): p. 99-103.
[31] Harper, J. et al., When and how should new technology be introduced into the IVF laboratory? *Hum Reprod*, 2012. 27(2): p. 303-13.
[32] Pribenszky, C. et al., Prediction of in-vitro developmental competence of early cleavage-stage mouse embryos with compact time-lapse equipment. *Reprod Biomed Online*, 2010. 20(3): p. 371-9.
[33] Hashimoto, S. et al., Selection of high-potential embryos by culture in poly(dimethylsiloxane) microwells and time-lapse imaging. *Fertil Steril*, 2012. 97(2): p. 332-7.

[34] Lemmen, J. G., Agerholm I. and Ziebe S., Kinetic markers of human embryo quality using time-lapse recordings of IVF/ICSI-fertilized oocytes. *Reprod Biomed Online*, 2008. 17(3): p. 385-91.
[35] Nakahara, T. et al., Evaluation of the safety of time-lapse observations for human embryos. *J Assist Reprod Genet*, 2010. 27(2-3): p. 93-6.
[36] Cruz, M. et al., Embryo quality, blastocyst and ongoing pregnancy rates in oocyte donation patients whose embryos were monitored by time-lapse imaging. *J Assist Reprod Genet*, 2011. 28(7): p. 569-73.
[37] Kirkegaard, K. et al., A randomized clinical trial comparing embryo culture in a conventional incubator with a time-lapse incubator. *J Assist Reprod Genet*, 2012. 29(6): p. 565-72.
[38] van den Bergh, M., Bertrand E. and Englert Y., Second polar body extrusion is highly predictive for oocyte fertilization as soon as 3 hr after intracytoplasmic sperm injection (ICSI). *J Assist Reprod Genet*, 1995. 12(4): p. 258-62.
[39] Mio, Y., Morphological Analysis of Human Embryonic Development Using Time-Lapse Cinematography. *J Mamm Ova Res*, 2006. 23(1): p. 27-35.
[40] Azzarello, A., Hoest T. and Mikkelsen A. L., The impact of pronuclei morphology and dynamicity on live birth outcome after time-lapse culture. *Hum Reprod*, 2012. 27(9): p. 2649-57.
[41] Scott, L. et al., The morphology of human pronuclear embryos is positively related to blastocyst development and implantation. *Hum Reprod*, 2000. 15(11): p. 2394-403.
[42] James, A. N. et al., The limited importance of pronuclear scoring of human zygotes. *Hum Reprod*, 2006. 21(6): p. 1599-604.
[43] Kiessling, A. A. et al., Evidence that human blastomere cleavage is under unique cell cycle control. *J Assist Reprod Genet*, 2009. 26(4): p. 187-95.
[44] Lundin, K., Bergh C. and Hardarson T., Early embryo cleavage is a strong indicator of embryo quality in human IVF. *Hum Reprod*, 2001. 16(12): p. 2652-7.
[45] Fenwick, J., et al., Time from insemination to first cleavage predicts developmental competence of human preimplantation embryos *in vitro*. *Hum Reprod*, 2002. 17(2): p. 407-12.
[46] Arav, A. et al., Prediction of embryonic developmental competence by time-lapse observation and 'shortest-half' analysis. *Reprod Biomed Online*, 2008. 17(5): p. 669-75.

[47] Meseguer, M. et al., The use of morphokinetics as a predictor of embryo implantation. *Hum Reprod,* 2011. 26(10): p. 2658-71.
[48] Wang, Y. L., The mechanism of cytokinesis: reconsideration and reconciliation. *Cell Struct Funct,* 2001. 26(6): p. 633-8.
[49] Koyama, H. et al., A high-resolution shape fitting and simulation demonstrated equatorial cell surface softening during cytokinesis and its promotive role in cytokinesis. *PLoS One,* 2012. 7(2): p. e31607.
[50] Rienzi, L. et al., Relationship between embryo morphodinamic and molecular karyotype in poorprognosis patients. *Hum Reprod,* 2012. 27 (suppl 2): ii286-ii302.
[51] Meseguer, M. et al., Embryo incubation and selection in a time-lapse monitoring system improves pregnancy outcome compared with a standard incubator: a retrospective study. *Fertil Steril,* 2012. Article in press.
[52] Hlinka, D. et al., Time-lapse cleavage rating predicts human embryo viability. *Physiol Res,* 2012.

INDEX

A

Abraham, 64
access, 2
ACF, 132, 133
acid, 49, 59, 68, 75, 84, 89, 96, 107, 109
adaptation, 22, 26, 27, 33, 36, 38
adhesion, 56, 74, 85, 101
adipocyte, 21
adipose, 21
adipose tissue, 21
adjustment, 36
advancement(s), 75, 96
age, 7, 8, 9, 14, 15, 17, 20, 21, 22, 25, 27, 30, 38, 59, 61, 129
allantois, 3
allele, 86
alters, 30
amino, 49, 59, 63, 67, 75
amino acid, 49, 59, 63, 67, 75
ammonia, 13, 15
amnion, 3
amphibia, 29
amphibians, 11
amputation, 96, 111
anatomy, 98, 110
androgen, 30, 73, 105
androgens, 10, 31, 36, 37
aneuploidy, 61, 134, 136
angiogenesis, 94
antagonism, 102
antigen, 60
apical ectodermal ridge (AER), vii, viii, 78, 99, 100, 103, 104, 105, 106, 108, 110, 111
apoptosis, ix, 78, 88, 91, 94, 97, 108
arginine, 49, 66, 70
Arid3b, ix, 78, 83, 85, 101
arrest, 59, 90, 115, 125
assessment, 59, 60, 61, 62, 63, 65, 136
assimilation, 14, 16
atmosphere, 114, 135
attachment, 42, 57, 73
attractant, 80, 85
avian, 3, 7, 10, 11, 23, 28, 31, 32, 36, 37, 38

B

basal lamina, 51
basic research, 58, 128
behaviors, 98
Beijing, 41
beneficial effect, 26
BIA, 97
bioavailability, 93
biochemistry, 39
biological processes, 79
biomarkers, 62
biopsy, 61, 76
biosynthesis, 52

Index

birds, 5, 6, 7, 10, 11, 12, 23, 28, 31, 34, 36, 37, 38, 104
births, 95
birthweight, 137
blastocyst, vii, viii, ix, 41, 42, 43, 44, 47, 48, 49, 51, 52, 53, 54, 55, 56, 57, 58, 59, 60, 61, 63, 64, 65, 66, 67, 70, 71, 72, 73, 74, 75, 114, 121, 126, 127, 128, 129, 130, 131, 132, 134, 135, 136, 137, 138
blastoderm, 13, 14, 28, 36
blastulation, ix, 113, 115, 137
blood, 29, 39, 66, 91
blood vessels, 91
bloodstream, 10
body size, 46
body weight, 8, 9, 14, 17, 18, 19, 20, 24, 29, 33, 36
bone, ix, 78, 81, 100
brachydactyly, 96
brain, 7, 22, 27, 28, 33, 54, 75
branching, 99
breeding, 44, 66
budding, 79, 80, 98

C

calcium, 56, 72, 74
CAM, 29
cancer, 3
carbohydrate, 72
carbohydrate metabolism, 72
carbon, 38, 135
carbon dioxide, 38, 135
carnivores, 44, 45, 46, 47, 52
carotenoids, 37
cartilage, 79, 84
cell biology, 101, 103
cell cycle, 47, 126, 129, 135, 138
cell death, 84, 88, 91, 92, 98, 100, 108
cell differentiation, 57, 94
cell division, 48, 75, 126
cell line, 42
cell movement, 101
cell organelles, 61
cell signaling, 81

cell surface, 51, 127, 139
central nervous system, 96, 106
chicken, vii, 2, 16, 18, 25, 27, 28, 29, 30, 31, 32, 33, 34, 37, 39, 105, 108, 111
China, 41, 62
chorion, 3
chromosome, 9
classification, 129
cleavage, 30, 58, 60, 113, 115, 122, 123, 124, 126, 127, 128, 129, 131, 133, 134, 136, 137, 138, 139
closure, 101, 104
clubfoot, 96, 110
colonization, 3
color, 20
commercial, vii, 1, 7, 9, 12, 15, 17, 18, 23
community, 89
compaction, 82, 115, 122, 130
comparative analysis, 36
compensatory effect, 57
competition, 7, 9, 10
complexity, 73, 105
complications, 68, 114
composition, 32, 116
conception, 73, 136
conditioning, 34, 38
conductance, 5, 9
congenital malformations, 96
consensus, 84, 90, 136
consumption, 50, 51, 59, 65
contraceptives, viii, 42
control condition, 19, 21
controversial, 28, 89, 90
coordination, 78
corpus luteum, 45, 46, 65
correlation, 115, 125, 135
cortex, 10
critical period, 11, 22, 24, 25, 33, 38
crossing over, 122
cryopreservation, 116
cues, 64
cultivation, 114, 115, 117, 118, 120, 126, 134, 135
cultivation conditions, 114, 116

Index

culture, 48, 49, 53, 60, 61, 65, 66, 72, 90, 94, 114, 115, 118, 119, 120, 134, 135, 137, 138
culture conditions, 48, 114, 115, 118, 134
culture media, 66, 72
culture medium, 49, 60, 118
cycles, 73, 116, 129, 137
cyclooxygenase, 52, 64
cytochrome, 33, 50, 70
cytokinesis, 115, 122, 124, 126, 127, 128, 129, 131, 133, 134, 139
cytoskeleton, 60, 134

D

defects, 83, 87, 93, 95, 96, 104
deficiency, 72, 75
Denmark, 118, 119
deposition, vii, 1, 27, 30, 34
deprivation, 70
depth, 61
detectable, 81, 90
detection, 50, 60, 90
developmental process, vii, 1, 4, 13, 42, 94, 95, 96
dialogues, 62
diet, 6
diffusion, 109
diplotene, 122
distribution, 26, 85, 107, 134
divergence, 109
diversification, 109
diversity, 7, 50, 90, 94, 95, 97, 110
DNA, 22, 28, 48, 56, 61, 66, 67, 73, 75, 137
DNA damage, 73, 137
DNA polymerase, 75, 137
dogs, 32
domestic fowl, vii, 1, 3, 7, 9, 26, 30, 37, 38
dopamine, 44, 69
dopamine agonist, 44
dosage, 9, 36
down-regulation, 56
Drosophila, 82, 99, 100, 104

E

E-cadherin, 64
ectoderm, 80, 81, 82, 84, 85, 86, 87, 89, 92, 100, 104, 105, 108
editors, 34, 36, 38
egg, viii, 2, 3, 4, 5, 6, 8, 9, 10, 11, 13, 14, 15, 16, 18, 19, 26, 27, 30, 36, 37, 38, 114
egg storage, viii, 2, 3, 4, 5, 13, 14, 15, 16, 26, 27, 30
electron, 51, 58, 63, 80
electron microscopy, 63
embryo development, vii, viii, ix, 2, 3, 4, 5, 13, 15, 17, 18, 25, 26, 27, 29, 42, 56, 59, 62, 67, 75, 113, 114, 115, 116, 117, 122, 123, 125, 126, 127, 128, 134, 136
embryo implantation, viii, 41, 45, 53, 55, 56, 57, 60, 61, 71, 72, 74, 75, 139
embryo transfer, vii, 52, 58, 59, 60, 63, 66, 68, 72, 114, 115, 136, 137
embryogenesis, vii, viii, 1, 2, 3, 5, 10, 11, 15, 16, 19, 21, 22, 23, 24, 25, 27, 29, 35, 38, 100, 104, 105, 107
embryonic stem cells, 50
endocrine, 6, 11, 28, 35, 44, 66
endocrinology, 34
endoderm, 104
endothermic, 25
energy, 3, 7, 8, 21, 23, 25, 27, 49, 51, 56, 59
energy expenditure, 3, 21
England, 98, 99, 100, 101, 102, 103, 105, 106, 109, 110, 111
environment, vii, 1, 3, 4, 6, 7, 12, 24, 33, 38, 45, 69, 118, 120
environmental change, 12
environmental conditions, vii, 1, 3, 5, 10, 12, 13, 18, 26, 44, 45
environmental effects, viii, 2, 5
environmental factors, vii, viii, 2, 4, 9, 13, 22
environmental influences, 7
environmental stimuli, 7, 44
environmental temperatures, 15, 23
environmental variables, 6, 10
enzyme, 9, 37, 50, 52, 53, 64, 90, 96

epigenetic modification, 12
epigenetics, 22
epithelial cells, 57, 64, 101
epithelium, 42, 51, 57, 62
equipment, 18, 114, 117, 120, 121, 137
estrogen, 10, 11, 45, 47, 48, 49, 51, 52, 53, 56, 67, 70, 75
ethical issues, 120
ethics, 58
Europe, 135
evaporation, 14
evidence, 13, 14, 20, 42, 55, 57, 59, 89, 91, 104, 105, 115
evolution, 35, 36, 94, 95
excision, 105
excretion, 24
exposure, 6, 7, 9, 11, 22, 24, 25, 26, 44, 49, 104, 116, 120, 131, 137
extracellular matrix, 94
extrusion, 122, 125, 130, 133, 135, 138

F

fat, 20, 21
femur, 95
fertility, 36, 58, 64, 74
fertilization, vii, ix, 4, 9, 13, 16, 44, 58, 72, 113, 114, 117, 123, 125, 126, 133, 136, 138
fetal development, 136
fetal viability, 61
fibers, 85
fibroblast growth factor, ix, 78, 81, 106
fibroblasts, 85
fibula, 95
films, 116
flank, 79, 98
Flrt3, ix, 78, 86
fluid, 37
follicle, 4, 59
follicle stimulating hormone, 59
food, 10, 12, 23
force, 127
Ford, 36

formation, 13, 16, 21, 23, 30, 55, 70, 78, 79, 81, 82, 84, 85, 86, 88, 90, 92, 94, 96, 98, 99, 100, 109, 115, 117, 120, 122, 127, 130, 134, 135, 137
fragments, 115, 117, 127, 134
fusion, 96, 104, 117

G

gastrula, 16
gastrulation, vii, 1, 106
gel, 57
gene expression, 2, 43, 56, 57, 66, 71, 80, 85, 87, 88, 91, 95, 102, 128
gene silencing, 39
genes, 9, 10, 23, 39, 55, 56, 64, 67, 72, 82, 84, 85, 89, 91, 92, 95, 98, 100, 101, 103, 106, 108, 127, 134
genetic background, 73
genetic defect, 128
genetics, 29, 98, 99, 102, 105, 106, 107, 111
genome, 34, 74, 94, 101, 108, 127, 137
genomics, 62
genotype, 12
germ cells, 10
gestation, 43, 45, 52, 64, 65, 67, 68, 69
gizzard, 14
gland, 45
glucose, 29, 49, 59, 63, 66, 70
glycerol, 54
glycogen, 20, 29, 51
glycolysis, 50, 70
gonads, 10, 34, 37
grading, ix, 59, 71, 114, 115, 137
granules, 51
growth, viii, 2, 3, 4, 5, 7, 9, 11, 14, 16, 17, 19, 20, 21, 23, 24, 25, 27, 28, 31, 32, 33, 34, 36, 38, 44, 45, 47, 52, 63, 64, 67, 68, 70, 71, 73, 74, 75, 87, 94, 95, 99, 100, 102, 103, 105, 107, 109
growth factor, 52, 63, 64, 67, 68, 70, 71, 73, 74, 75, 99, 105
growth rate, 95, 109
guidance, 94

H

habitat, 3, 31
hair, 102
hair follicle, 102
harmful effects, 13
health, 30, 61, 74
heart rate, 25
heat loss, 23
heat shock protein, 38
height, viii, 2, 4, 9, 14, 28, 35
heredity, 33
heterochromatin, 51
history, 36, 95, 98
HLA, 60, 70
hormone levels, 31
hormones, vii, 2, 4, 5, 6, 7, 10, 11, 12, 14, 15, 16, 24, 25, 28, 30, 31, 33, 34, 44, 46, 53, 66
housing, 7
human, ix, 50, 57, 58, 59, 60, 61, 62, 63, 64, 65, 66, 67, 68, 71, 72, 73, 75, 94, 95, 96, 97, 101, 102, 105, 110, 111, 113, 114, 116, 117, 120, 126, 129, 135, 136, 137, 138, 139
human genome, 126
human leukocyte antigen, 72
humidity, 4, 13, 22, 33
Hungary, 117, 118
H-Y antigen, 29
hybridization, 61, 65
hydrolysis, 70
hyperplasia, 91
hyperthermia, 23, 25
hypertrophy, 21, 35
hypothalamus, 22, 24, 25
hypothesis, 24, 60, 68, 89
hypoxia, 4, 50
hypoxia-inducible factor, 50

I

identification, 43
identity, 39, 108
idiopathic, 129
illumination, ix, 113, 117, 119
image, 79, 80, 81, 96, 111, 117, 118, 119, 120, 121, 131
immune system, 3
immunity, 37
immunoprecipitation, 94
implantation, vii, viii, 41, 42, 43, 44, 45, 47, 48, 49, 50, 51, 52, 53, 54, 55, 56, 57, 58, 59, 60, 61, 63, 64, 65, 66, 67, 68, 69, 70, 71, 72, 73, 74, 75, 87, 93, 107, 115, 116, 121, 123, 124, 128, 129, 135, 136, 138
imprinting, 116
in situ hybridization, 61, 81
in vitro, vii, ix, 48, 49, 53, 58, 63, 66, 70, 71, 72, 85, 113, 114, 115, 117, 120, 128, 135, 136, 137, 138
in vivo, 3, 48, 49, 85, 90, 103, 137
incidence, 61
incisor, 109
incubation, vii, viii, 1, 2, 3, 4, 5, 7, 9, 11, 12, 13, 15, 16, 17, 18, 19, 21, 22, 23, 24, 25, 26, 27, 28, 30, 32, 33, 34, 35, 36, 39, 139
incubation period, vii, viii, 1, 2, 5
incubator, 16, 18, 34, 117, 118, 119, 120, 121, 129, 131, 137, 138, 139
incubators, ix, 17, 23, 113
indentation, 134
individuals, 96, 110
induction, viii, 23, 38, 46, 52, 57, 69, 78, 82, 83, 86, 96, 99, 100, 105, 107
industry, 9, 116
infertility, viii, 42, 129
infrared spectroscopy, 72
inheritance, 34
inhibition, 44, 45, 53, 55, 87
inhibitor, 53
initiation, 5, 13, 15, 26, 42, 46, 47, 57, 70, 79, 80, 81, 86, 92, 94, 97, 98, 99, 106, 123, 124
injections, 3, 47
inositol, 72
insulation, 12, 23
integration, 24
integrins, 85

integrity, 86
integument, 100
intracellular calcium, 74
investment, 33
Israel, 1, 27
issues, 13, 58

K

karyotype, 128, 139
keratinocytes, 101
kidney, 10, 95, 102
kinetic parameters, 127, 128, 135
kinetics, 131

L

lactate dehydrogenase, 50, 56
lactation, 43, 44, 45, 46
larvae, 96
lead, 7, 89, 115, 127, 134
LED, 117, 119
legs, 96
leucine, 49, 70, 104
leukemia, 52, 71, 72
life cycle, 3, 7
ligand, 57, 60, 64, 70, 82, 86, 92, 93
light, 25, 44, 69, 117, 118, 119, 120, 131, 137
limb buds, viii, 77, 79, 87, 89, 92, 94, 96, 97, 102, 108, 109
lipids, 37
liver, 14, 20, 28
longevity, 51
low temperatures, 11, 23
lumen, 51
luteinizing hormone, 125
lymphocytes, 68
lysosome, 58

M

majority, 17, 44, 115, 121
malate dehydrogenase, 50

male bias, 4
mammals, 9, 10, 42, 43, 44, 45, 52, 71, 109
management, 4, 5, 7, 9, 11, 16, 26, 29
manipulation, 10, 19, 21, 29, 30, 31, 37, 61, 87, 88, 90, 93, 94
mapping, 89, 108
marketing, 17, 20, 21, 23, 24
mass, 6, 11, 42, 48, 50, 51, 54, 58, 59, 62, 66, 67, 71, 115, 122, 126
maternal effects, vii, viii, 1, 2, 3, 4, 5, 6, 27, 31, 36
matter, 31
maturation process, 82, 85
measurement, 23, 32, 36, 63, 66
meat, 3
media, 65, 114, 116, 118, 120
median, 124, 131
medicine, 120
medulla, 10
meiosis, 4, 114, 122, 125
melatonin, 44, 46, 69
membranes, 3, 55
memory, 28
mesenchyme, 80, 81, 82, 84, 85, 87, 88, 89, 91, 92, 93
mesoderm, viii, 10, 77, 79, 98
messenger RNA, 63
metabolic, 37, 58
metabolic pathways, 56
metabolism, 13, 30, 33, 34, 48, 50, 51, 56, 59, 66, 67, 68, 76, 116
metabolites, 49, 53, 59, 60
metaphase, 125
metastasis, 101
metatarsal, 95
methylation, 22, 28, 61
mice, 42, 43, 47, 48, 49, 52, 53, 54, 55, 57, 61, 64, 65, 69, 75, 81, 83, 89, 91, 92, 100, 110, 111, 126
micrograms, 80
microRNA, 108
microscope, 51, 58, 117, 118, 119, 135
migration, 79, 85, 94, 101, 104
miniature, 117
Minneapolis, 77

miscarriage, 66, 68
Missouri, 30, 31
mitochondria, 50, 61
mitosis, 124
model system, ix, 43, 78, 94
models, viii, 2, 3, 29, 41, 43, 55, 58, 61, 68, 74, 78, 88, 89, 95, 96, 98, 109, 114, 121
modifications, 7, 12, 33
molecules, viii, ix, 41, 43, 52, 56, 57, 61, 78, 81, 91, 93
Moon, 106
morphogenesis, viii, 77, 78, 79, 93, 98, 99, 101, 103, 105, 106, 109
morphology, 6, 30, 58, 60, 63, 95, 97, 114, 115, 116, 121, 125, 138
morphometric, 109
mortality, 5, 12, 14, 15, 16, 23, 25, 26, 29, 38
mortality rate, 26
morula, 59, 67, 122, 130
mother cell, 127
motif, 60
mRNA, 34, 53, 54, 73
multilayered structure, 83
mutant, 57, 73, 83, 85, 90, 91, 100
mutation, 90, 110, 111
myogenesis, 106

N

National Academy of Sciences, 70, 99, 101, 103, 104, 108, 112
National Institutes of Health, 97
natural selection, 115
negative effects, 120
neonates, 110
nerve, 94
nervous system, 96
Netherlands, 36
neurodegenerative disorders, 61, 76
neutral, 42
NIR, 58, 60
nuclei, 21
nucleolus, 48, 49
null, 57

nutrient, vii, 2, 4, 5, 7, 14, 15, 16, 28, 65
nutrient intake, vii, 2, 5, 7, 28
nutrition, 5, 7, 8, 30, 36
nutritional status, 26

O

oocyte, 74, 121, 125, 126, 128, 134, 135, 137, 138
opportunities, 38, 93, 94
orbit, 49
organelles, 61
organize, 11
organs, 20, 27, 29, 33, 94, 95, 96, 100
ovariectomy, 48, 69
oviduct, vii, 1, 13, 16, 54
oviposition, vii, 1, 4, 13, 14, 16
ovum, 137
oxidative stress, 58, 60
oxygen, 25, 33, 70, 114, 135
oxygen consumption, 25, 70

P

parallel, 3
parents, 12
participants, 63
pathology, 100
pathways, 7, 50, 66, 91, 98, 99, 127
PCR, 61, 127
penalties, 17
periimplantation, viii, 42, 53, 57, 64, 69, 70
perinatal, 34
personal communication, 14
PGD, 52, 61
PGE, 52
phalanx, 96, 108
phenotype, 4, 6, 11, 12, 31, 32, 73, 83, 110
phenotypes, 7, 11, 30, 83, 98
phenotypic variations, 100
phosphorylation, 50, 93
physicians, 58
physiology, 6, 29, 65, 70
pigs, 44

pilot study, 72
pineal gland, 44
plant growth, 116
plasticity, 6, 12, 32
platform, vii, 79, 93, 97
playing, 6
ploidy, 65, 128
point mutation, 110
polar, 48, 51, 65, 122, 123, 127, 130, 133, 135, 138
polar body, 65, 122, 123, 130, 133, 135, 138
polarity, 99
policy, 114, 116, 137
polydactyly, 83, 87, 93, 96, 109, 110, 111
polymerase, 48
polypeptides, 105
population, 37, 58
porosity, 14
Portugal, 77, 97
position effect, 37
poultry, 7, 9, 22, 30, 38
precursor cells, 81, 82, 85, 92, 94
predators, 12, 23
pregnancy, 42, 43, 48, 49, 53, 54, 55, 57, 58, 59, 60, 61, 63, 65, 66, 67, 68, 69, 71, 72, 75, 114, 115, 121, 126, 129, 134, 136, 138, 139
preparation, 55, 120
prevention, 88
probability, 128, 129
progenitor cells, 111
progeny, vii, 2, 4, 5, 6, 8, 9, 12, 14, 16, 23, 26, 28, 36
progesterone, 46, 48, 49, 53, 65, 73, 75
prognosis, 136
programming, 3, 5
prolactin, 44, 45, 46, 47, 69, 70, 71
proliferation, 21, 27, 35, 47, 48, 51, 56, 63, 79, 85, 89, 101
prophase, 122
proposition, 65
proteasome, 60
protein kinase C, 57
protein synthesis, 47, 49, 51
proteins, ix, 23, 37, 78, 86, 134

proteomics, 62, 67
psychological stress, 116
pullet rearing phase, vii, 2, 5, 7

R

radiation, 23, 26
radius, 79, 95
Raman spectroscopy, 58, 72
reagents, 103
receptors, 52, 53, 54, 57, 64, 68, 74, 75, 86, 103
recombination, 122
reconciliation, 139
redundancy, 78, 82, 92
regenerate, 96
regeneration, 94, 96, 111
regression, viii, 78, 80, 87, 88, 97
relativity, 59
relief, 119
replication, 56
reproduction, 28, 65
reptile, 36
requirements, 13
researchers, 16, 94
resistance, 37
resolution, 118, 119, 139
resources, 6
response, 6, 11, 13, 22, 24, 26, 27, 31, 36, 45, 47, 56, 69, 89, 103, 104, 107, 125
responsiveness, 11
restoration, 96
reticulum, 51
retinol, 90
ribosome, 48, 51
risk, 61, 76, 120
RNA, 47, 48, 63, 67, 69, 94
RNAs, 39
rodents, 43, 44, 45
R-Spondin 2, ix, 78

S

safety, 117, 120, 121, 138

Index

scope, 118
secondary sexual characteristics, viii, 2, 4, 5, 12, 17
secrete, 60
secretion, 10, 44, 45, 46, 47, 48, 59
sensitivity, 38, 75, 90, 117, 120, 128
sensors, 117
serum, 65
sex, 4, 6, 9, 10, 11, 12, 27, 29, 30, 31, 33, 35, 36, 37, 38, 39
sex chromosome, 4, 9
sex differences, 12
sex ratio, 4, 11, 27, 31, 36, 38
sex reversal, 10, 11, 36
sex steroid, 10, 11, 30
sexual behavior, 11
shape, 62, 71, 80, 98, 127, 139
shear, 31
shear strength, 31
shock, 16, 17, 23, 26
showing, 52, 55, 59
sibling, 135
signal transduction, 104
signaling pathway, 74, 86, 96, 98, 102, 134
signalling, 75, 102, 103, 104, 105, 106, 109, 127
signals, 6, 80, 84, 89, 99, 100, 102, 106, 107, 108, 127
simulation, 139
skeletal features, 79
skeletal muscle, 35
skeleton, 79, 83, 88
skin, 95
social influence, 35
social stress, 44
software, ix, 113, 131
solution, 116
Spain, 77, 97
species, ix, 2, 6, 10, 11, 27, 31, 32, 42, 43, 44, 46, 47, 52, 60, 79, 90, 95, 97, 113, 120
sperm, 61, 114, 117, 125, 126, 128, 137, 138
spindle, 134
spleen, 54

spontaneous pregnancy, 42, 54
Spring, 104, 107, 110
standard deviation, 132, 134
stars, 116
state, 3, 42, 45, 47, 70
stem cells, 50, 111
steroids, 6, 10, 11, 12, 44, 46
stimulus, 45, 46
storage, viii, 2, 3, 4, 5, 13, 14, 15, 16, 17, 18, 26, 27, 28, 29, 30, 31, 36, 38
stress, 7, 10, 23, 24, 26, 34, 37, 38, 44, 45, 60, 85, 116
structure, 22, 48, 78, 79, 80, 82, 85, 86, 88, 95, 96, 97
subgroups, 134
subjectivity, 116
substrate, 3, 11
success rate, 58, 134
sulfate, 57
Sun, 101, 105, 106, 108
supplementation, 29, 49
surface area, 5
surgical removal, 88
survival, 6, 20, 29, 36, 37, 44, 45, 68, 89, 101
swelling, 51
Switzerland, 34, 76
synchronization, 42
synchronize, 54
syndrome, 110
synthesis, 10, 24, 48, 49, 63, 66, 67, 70, 90, 107

T

target, 70, 85, 102
techniques, 114
technology, 7, 115, 116, 120, 121, 122, 128, 129, 135, 137
temperature, viii, 2, 4, 5, 11, 13, 14, 15, 16, 17, 18, 19, 20, 22, 23, 24, 25, 26, 27, 28, 29, 30, 31, 32, 33, 36, 37, 38
teratogen, 104
testing, 82, 120, 121
testis, 10

testosterone, 4, 10, 17, 36, 37, 38
thermoregulation, 24, 34, 35
thoughts, 89
thyroid, 22, 24, 25, 32, 35
thyroid gland, 24, 25, 32
thyroid stimulating hormone (TSH), 24
tibia, 95
time frame, 115, 126
time periods, 42
time variables, 128
tissue, 3, 21, 23, 29, 80, 87, 94, 96, 111
tooth, 106
trafficking, 74
training, 120
traits, vii, 2, 3, 4, 5, 6, 39
transcription, 52, 83, 85, 89, 91, 96, 101, 102, 107, 110
transcription factors, 83, 85, 91, 101, 107
transcripts, 104, 128
transducer, 93
transduction, 23, 107
transformation, 10, 37
transgene, 86, 90, 107
translation, 48
translocation, 55
transmission, 51
transplantation, 3
transport, 54, 74
treatment, 15, 16, 17, 18, 19, 21, 25, 44, 47, 55, 59, 62, 134
trial, 121, 129, 138
tricarboxylic acid, 50
tricarboxylic acid cycle, 50
triggers, 105
triiodothyronine, 29, 32
tumor, 83
tumors, 103
Turkey, 28, 32
turnover, 58, 59, 63, 67

U

ubiquitin, 60
ulna, 79, 95

ultrastructure, 71
uniform, 12, 16, 23
united, 27, 70, 99, 101, 103, 104, 108, 112
United States (USA), 27, 70, 99, 101, 103, 104, 108, 112, 137
uterine receptivity, viii, 41, 43, 53, 54, 61, 68, 71, 73
uterus, 42, 52, 53, 54, 55, 56, 64, 67, 70, 71, 72

V

variations, 17, 34, 61, 134
vascularization, 15
velocity, 16
vertebrates, 34, 69, 78, 88
vesicle, 74
videos, 117

W

walking, 9
wallabies, 47
water, 3, 14, 37
weight loss, 38
wind speed, 37
Wnt signaling, 55, 63, 73, 103
worldwide, 17
wound healing, 96

Y

yield, 32
yolk, vii, 1, 4, 5, 6, 7, 9, 10, 12, 14, 15, 16, 20, 21, 24, 26, 27, 31, 32, 34, 35, 36, 37
yolk hormones, vii, 2, 4, 5, 12

Z

zinc, 83, 101
zygote, ix, 114, 131